I0486414

Macrocreation

An Energy Explanation Of Existence

by

Kevin Christopher Pratt

Bloomington, IN  Milton Keynes, UK

AuthorHouse™
1663 Liberty Drive, Suite 200
Bloomington, IN 47403
www.authorhouse.com
Phone: 1-800-839-8640

AuthorHouse™ UK Ltd.
500 Avebury Boulevard
Central Milton Keynes, MK9 2BE
www.authorhouse.co.uk
Phone: 08001974150

© 2007 Kevin Christopher Pratt. All rights reserved.

No part of this book may be reproduced, stored in a retrieval system, or transmitted by any means without the written permission of the author.

First published by AuthorHouse 2/2/2007

ISBN: 978-1-4259-8180-8 (sc)

Library of Congress Control Number: 2006910759

Printed in the United States of America
Bloomington, Indiana

This book is printed on acid-free paper.

Table of Contents

Section One—Introduction

Opening Remarks

The central purpose of this book is to present the trinity of principles that constitute the macrocreation theory. This theory attempts to offer a logical and rational explanation of how all aspects of existence came into being. This book will provide the principles of the theory and then offer speculation on how these principles can help to explain everything in this reality of which we are a part. However, this book will never make the claim that the macrocreation theory is *truth*. Anyone who claims that their own personal belief system guarantees absolute truth is acting from a deep-seated psychological need to believe that claim. This book offers only ideas and speculation.

For a very long time, society has been engaged in the shopworn philosophical debate of religious creationism versus atheistic physics. This unproductive debate needs to come to a timely end. Although mainstream science has attempted to be rational in its theorizing regarding the origin of the universe, I believe that the scientific establishment has failed to provide a truly coherent hypothesis for existence. The time has come for our philosophical debate to evolve beyond the old and inadequate notions that religion and science have provided. The macrocreation theory is a fresh and comprehensive way to understand everything in our world. If it accomplishes nothing else, it will help shift our age-old philosophical questioning to a new level.

The macrocreation theory attempts to answer all the major questions that have perplexed humanity for its entire history. These questions will be answered within a framework that is internally consistent and that utilizes fundamental principles that are based on straightforward logic. But I will state once again that none of these characteristics in any way validates the macrocreation theory as being truth.

One thing that this theory does not attempt to do is to help people feel good about themselves or about life. Although I do not claim that the macrocreation theory is true, I do believe that truth —whatever truth is—should not be linked to anything else, including such things as love, benevolence, or spirituality. Truth needs to stand alone, separate, and apart from human emotions. To put it quite simply, truth is not human.

In the past, many thinkers have proposed the concept that the world that we live in is an imponderable mystery. This means that the basic proposition to consider is that either the world we live in can be explained, or it cannot. Throughout history, all religions have provided a jumble of explanations for the nature of our existence. Most people recognize that these "explanations" are, in fact, myths—designed by human beings to meet human expectations. These myths may appeal to human sensibilities but generally do not appeal to human logic.

Usurping the role of religions, the scientific elite have taken the lead role in trying to comprehend the truth of our existence. On the surface, at least, these scientific explanations appear superior to the myth-making of ancient religions. However, scientific theorizing has proven quite inadequate as well.

All that science has put forth so far are scattershot ideas that leave vast gaps in logic and coherence. For example, current theories state that much of the universe is composed of "dark matter" and "dark energy," yet no scientist can say, with any precision, what these "dark" things are. All that mainstream science can offer to a gullible public are frequently revised wild guesses. Rational human beings should not accept this state of affairs.

This book is organized into four sections. The first section provides a general philosophical context for the macrocreation theory. Because this theory is revolutionary in nature, it will be helpful to have an intellectual framework for it. The second section puts forth the trinity of principles, along with allied corollaries, that comprise the macrocreation theory itself. The third section applies, in an ambitiously speculative manner, those principles to virtually all aspects of life and existence. The speculative material that is provided in this section is, without question, inspired by my subjective viewpoints. All individuals may take the trinity of principles and pursue a philosophical course that is in keeping with their own unique perspectives. The fourth section of this book details my philosophical development and how I arrived at the conclusions that make up the macrocreation philosophy.

This book is intended to offer a *living* philosophy that will be forever open to new thinking and to new thinkers. My hope is that the macrocreation theory will open up fresh horizons for fruitful philosophical endeavors for generations to come. Perhaps the next great evolutionary step that humanity will experience will be a significant advancement in the comprehension of this reality that we are all a part of. If we can collectively come to understand why this world exists and how it functions, then this knowledge should help to free us from the disturbed confusion that has dominated human history thus far. Just possibly, logic and rational thinking may become the new order of the day.

This book is fairly short and succinct and will not drown the reader in an endless sea of words. Philosophical enlightenment—if it happens at all—comes from the effort of thought and not from the effort of reading. I believe that throwing too many words at the reader, especially when the subject matter is complex, can harm the ability of the reader to understand the concepts that are being conveyed. The many new ideas that are found in this book must be fully contemplated if they are to be fully understood. In other words, while the reading experience should be rather easy, the thinking experience may present a significant challenge.

Scientific Fundamentalism

Fundamentalism is a psychological condition that can have an impact in the supposedly rational and unemotional realm of science. This form of fundamentalism is as truly counterproductive to intellectual objectivity as is religious fundamentalism. All too often, scientific fundamentalists construct rigid mental barriers that serve to maintain a narrow framework of reality that does not stand up to truly objective scrutiny.

Fundamentalism of any type derives from a profoundly deep psychological need. Many people require a sharply defined mental zone that allows them to comfortably maneuver through life. They simply cannot deal with a wealth of theoretical possibilities being placed in front of them. It is very important to this psychological type that their particular belief system involves only *settled* issues. This means that they do not have to continually reevaluate any new information that may come their way.

In the case of scientific fundamentalists, this group understandably rejects religious philosophies due to these systems' quite obvious failings in logic and rationality. With the rejection of theologies, scientific fundamentalists then embrace the ideas of the conservative scientific elite. Once embraced, scientific fundamentalists are as loathe to reconsider their core beliefs as are religious fundamentalists. Scientific texts become as sacred as cherished sacred texts. It should be stated that neither scientific fundamentalists nor religious fundamentalists have inadequate intellects, they are simply overwhelmed by their emotional insecurities.

Fundamentalists of any type are generally quite submissive to their own authority figures. For the religious faithful, obviously, it will be the church hierarchy. For the scientific faithful, it will be the highly credentialed members of the scientific establishment. Fundamentalists place their faith in these authority figures for reasons relating to deep personal insecurity. This high level of anxiety results in a lockbox mentality, with autocratic authority figures tightly clutching the keys.

As a rule, most human beings do not want to bear the burden of constantly reevaluating the belief system that frames their reality. This fact is certainly understandable. People simply wish to embrace a system of belief that sounds reasonable to them and then go on with the hectic business of living their lives. However, for neurotic personal reasons, fundamentalists of any type feel the need to bolt the doors against any new ideas that may take shape around them. This intense need leads to a closed mind, and a closed mind is an irrational mind.

In contrast to religious fundamentalism, the nature of scientific fundamentalism has not come under much scrutiny by society at large. Instead, a rigid scientific mindset has been considered the paragon of human rationality. Healthy intellectual evolution requires that the steel walls of scientific fundamentalism be torn down by those who recognize the fact that only an open mind can be a genuinely rational mind. Society at large needs to turn away from those figures of scientific or religious authority who would presume to explain this world of ours in terms that correspond to their individual insecurities. For the sake of our philosophical evolution, we all must think for ourselves.

The Role of Science in Society

The scientific establishment in today's society mirrors the exalted role of the church hierarchy in Europe during the Middle Ages. Then, as now, these respected authorities have taken it upon themselves to provide official explanations to the public of the nature of the world around them. Sadly, to a significant extent, most people choose to unthinkingly embrace these philosophical edicts as they are issued. In a very real sense, we have not actually come that far from the Dark Ages of long ago.

It needs to be said, however, that scientists are serving this domineering role in society because it has been useful for them to do so. Even many of the die-hard religious faithful who populate this world can no longer fully subscribe to the obviously mythical information that fills their sacred texts. Religious leaders can no longer be taken entirely seriously when it comes to providing explanations for the origin and nature of our existence. Therefore, the scientific establishment has zealously seized this role for themselves.

Society now expects our scientific authorities to fulfill their function—to stand and deliver plausible-sounding explanations for all the perplexing universal questions. Society has generously given the scientific elite great latitude in meeting their societal obligations. Each new pronouncement on such topics as the purported "age" of the universe is taken seriously by the public because people have a psychological impetus to take these pronouncements seriously.

The present-day "religion" of mainstream science has become increasingly dogmatic and self-protective. Those who dare to question the edicts that are issued from "on high" are quickly labeled heretics and soon become marginalized. The frequent nonsensical goofiness of the New Age movement has served to bolster the conservative gray eminence displayed by the scientific authorities. The philosophical status quo has thus become quite stagnant.

The really "big" questions relating to the origin and nature of existence are not subject to being answered using standard scientific techniques employed in a laboratory setting. Superficial observation of physical phenomena will only get you so far, and no farther. What the scientific establishment has been providing to society for too long a time has been merely pseudo-rational concepts in the realm of physics. These inadequate concepts deserve to be put under greater scrutiny.

Scientists who continually offer an endless series of recklessly wild guesses regarding such topics as the age or origin of the universe should not be taken very seriously by society at large. The faith that people have placed in the scientific elite has too often been misplaced. When it comes to the major philosophical questions that frame our existence, authority should never be invested in anyone. We all must think things through for ourselves, even if this means a lot of hard work. Hopefully, this book will prove helpful in that effort.

The macrocreation philosophy offers an absolute minimum of doctrine. Beyond the trinity of principles itself, all other aspects of the philosophy are open to individual interpretation and analysis. This book provides my personal interpretation and analysis primarily as a means of opening philosophical doors that were never meant to be closed. Your analysis has to be considered as philosophically valid as mine or anyone else's. Authority remains vested in you, and you alone.

Faith and the Lack of Faith

Despite what most people believe, the function of the world's religions has never been to provide truth to the populace. The purpose of religion has always been to provide solace, social support, and meaningful myths to society. These myths have not been mere trifles but have been of supreme importance to the development of human culture. Theology *is* mythology. Religions have provided a wealth of vibrant mythological frameworks that have creatively shaped the direction of human society.

The faith that an individual places in a religious framework can be extremely beneficial, and this benefit should not be minimized. At their best, religions provide social solidarity, guidance, compassion, and a benevolent spirit. For many people, these attributes are essential to their lives. The fact that the world's religions do not offer the full degree of truth that they claim to offer is not, in fact, a fatal flaw. Since it is not the purpose of all the world's theologies to be a source for truth, the failure of these theologies to provide truth is nearly irrelevant. Religions have always functioned in the manner that they needed to function. Nothing has gone amiss.

In contrast, the macrocreation theory has no compassion or benevolence to offer. There is nothing intrinsically "feel-good" about logic and rationality. The macrocreation theory is not human-centric— it has no relation to human desires or aspirations. While this book certainly serves as an advocate for the macrocreation philosophy, I am not an evangelist. I have no interest in attempting to convert the religious faithful to the ideas of the macrocreation belief system. Those people who are satisfied with their religious framework should probably not even consider abandoning the church or temple or mosque that has provided them with the sustenance that they need.

There are others, however—atheists, agnostics, and independent thinkers—who possibly may benefit from the ideas of the macrocreation philosophy. Neither scientific fundamentalism nor the New Age movement has provided much useful guidance in explaining the nature of this world. It can be stated with some assurance that there is room for, and that there is need for, a revolutionary philosophical system to take its place beside those that are already well established. Many people find this world a place of rampant confusion and simply do not know what to think. This new philosophical structure gives them a unique additional option.

Despite the apparent diversity that the world's religions display, it is not intellectually unreasonable to consider them all as one group. There can be no doubt that the theologies of Hinduism and Islam, for instance, present a wealth of philosophical differences. These variances are numerous and quite distinct and have been the basis for many years of conflict among their adherents. Yet the essential fact remains that the theologies of all religions—no matter what their various doctrinal characteristics may be—lack a genuine underlying rationality. For this reason, all the religions of the world can be considered as one because they all offer non-rational philosophical systems.

The purpose of this book is not to provide a thorough examination of the failings of the logic of any or all of the religions of the world. Thousands of theologies—ranging from the rigid and doctrinaire to the open and thoughtful—are all variations of imaginative mythmaking. None of them can be taken seriously on a purely logical basis. Religions can have a wonderful abundance of truly helpful qualities to offer a human being, but logic and rationality cannot be found among those qualities. Those beneficial qualities that any particular religion may have to offer will not be contemplated in this book. Thus, all religions will be treated as one.

I will reiterate my sincere statement that spiritual or religious faith is a magnificent thing and that no one should ever lightly walk away from this source of strength. If faith can be held on to, it should be held on to. Indeed, there are those who may make the argument that the ideas of the macrocreation theory can be used to "prove" the existence of "God." I will not be critical of those efforts even though they do not appeal to my personal sensibilities. I have no doubt that the established religions of the world will continue to play a powerful role in human society for many generations to come. Someday, though, cultural evolution may demand that a new intellectual era must begin to take shape.

New but Not New Age

People who consider themselves to be rationalists have been understandably dubious over the years about what the New Age movement has had to offer. Most people can recognize "mumbo jumbo" when they see it. However, simply because a plethora of nonsensical statements have been made in relation to a particular subject area does not invalidate in any way the subject area itself. For example, rationalists have seen all the many pie-in-the-sky descriptions of heaven made by dreamy-eyed religious types and have concluded from this that the concept of heaven is merely wishful thinking. The macrocreation philosophy will attempt to show that all the many fairytale conceptualizations that have been presented to describe the afterlife can be thrown out the window and that this subject area, along with numerous others, can be examined in a fresh, and rational, light.

It cannot be argued that, for thousands of years, the vast majority of humanity has felt an instinctive connection to something that is greater than themselves. It is fair to say that, for the most part, these people can be considered rational, coherent human beings. For all of human history, religions have taken the lead role in providing explanations for what this innate sense of connectedness actually is. In recent times, the New Age movement has offered its own set of generally quite vague, humanistic explanations. Thus, those people seeking to understand the nature of the link that they feel they have to this "divine" aspect of life have usually embraced a religion or New Age philosophy that seems to make some sort of sense to them. Religions and the New Age movement have loudly proclaimed that they possess the "truth," and so people have tended to listen to them. It must be said, however, that these forthright claims regarding the realm of "higher truth" are empty claims.

A scientific fundamentalist makes the contention that these people—billions of them—have all been delusional to a greater or lesser extent. An atheistic rationalist believes that there is no such thing as the "divine." Even though all the explanations of the "divine" that religions and New Age groups have had to offer have been thoroughly irrational, this does not mean that it is impossible to provide a new explanation for the "divine" that *is* rational. For scientific fundamentalists to assume that billions of otherwise rational people are delusional about this one thing is in itself an irrational conclusion.

The admirable goal of mainstream science—to conduct an objective search for rational truth—has been subverted by the defensive dogma of the scientific fundamentalists. These pretenders to rationalism have feverishly attempted to make their personal opinions the ultimate proclamations of "approved" rationality. A genuine search for rational truth cannot be conducted within the corrosive confines of any manner of fundamentalism. For the sake of cultural evolution and the liberation of philosophy, the scientific elite need to be removed from their pedestals of power.

The macrocreation theory is a belief system that exists outside the traditional boundaries of religion, the New Age movement, and mainstream science. The trinity of principles that comprise the macrocreation theory is founded, I believe, on a solid application of logic and rational deduction. These ideas may or may not be "true." There is no way to know for certain.

It should be reiterated that the macrocreation theory is not designed as a self-help tool in any way. These ideas may possibly assist a person in navigating a course through life, or they may not. The macrocreation philosophy is meant to be rational and coherent, but it is not necessarily useful as a source of comfort.

The next section of this book presents the trinity of principles that serves as the foundation for the macrocreation philosophy.

Section Two—The Trinity of Principles

Opening Remarks

The macrocreation theory is composed of a trinity of foundational principles. These principles can be applied, on a speculative basis, to every aspect of existence. These principles are quite straightforward and fairly simple to explain.

In addition to the trinity of principles, this section presents two short chapters that provide allied corollaries of deductive reasoning. These principles and corollaries have been laid out in a very succinct manner. Therefore, while the reading experience of this section will be brief, a certain amount of contemplation will be necessary in order for the reader to analyze the logic that was used in the formation of these ideas.

Some people may wish to make the effort to see a strong correlation of the trinity of principles of the macrocreation theory with the Christian Trinity. From my perspective, however, the Christian Trinity can be considered something comparable to a poetic endeavor rather than a precise formula for existence. This means that any correlation that can be seen between the Christian Trinity and the macrocreation theory will be mostly in the eyes of the beholder. I have made no effort to determine what possible relationship exists between these two trinities. This area is certainly open to exploration by anyone who may be interested in doing so.

The three principles of the macrocreation theory are foundational and irreducible. The macrocreation theory is intended to provide the basic formula for existence itself. The corollaries offered in this section concern the nature and function of time as well as attempt to explain why there is existence in the first place, as opposed to the valid alternative of nonexistence.

I believe that humanity can acquire coherent answers to all the questions that have intrigued us for our entire history as a species. The macrocreation theory is my effort, with all the best intentions, at the achievement of that ultimate goal of full understanding. This book is certainly not meant to be the last word. There will never be a last word. Philosophical insight, to be sure, will continue to evolve.

The First Principle

For generations now, physicists have worked hard to explain the composition of the universe and have come up with a variety of esoteric possibilities. One formula that has been proposed says that the universe is constituted of only a small percentage of "ordinary" matter, with everything else being distributed among the categories of energy, "dark energy," and "dark matter." This is quite an exotic concoction since physicists have been largely unable to explain the actual nature of these "dark" aspects of existence.

The macrocreation theory offers a much less convoluted formula. According to this conception, energy—in all its permutations—underlies everything in existence. Another way to state this is that energy is the sole substance of existence. It is an essential law of science that energy cannot be created or destroyed. This may very well mean that energy constitutes existence itself.

The belief that energy may be the universal substance of existence is certainly one that has been considered before. The main problem that some physicists have with this idea is that it presumably does not take into account the realities of atomic-level functioning. While it is indisputably an intellectual challenge to determine how the workings of atomic structures fit in with a universal energy essence, the admitted difficulty of this challenge should not be thought of as a discrediting factor. The answer may be hard to come by, but the solution is most likely there to be found.

Physicists and astronomers have tended to fix their attention upon all the dazzling displays of energy that the universe has to offer. These phenomena include black holes, novas, gravitational fields, curved space, and an endless sea of spiraling galaxies. These manifestations are certainly all quite fascinating. This mysterious universe can be rightfully considered mind-boggling in its complexity. It should be understood, however, that a contemplation of all the wondrous "light shows" that the universe has to offer may not necessarily lead to a fuller comprehension of the fundamental nature of existence.

Logic must be used to deduce the underlying essence of this world that we live in. We can say that energy is an irreducible element. It is the only aspect of existence that is irreducible. Therefore, everything around us (and inside of us) can be understood in terms of formulations of energy. Nothing lies outside this framework. Therefore, the first principle of the macrocreation theory is that energy, in all its permutations including atomic, is the sole substance of existence.

The Second Principle

There are some basic statements that can be made about energy. First of all, energy is never inactive. It is always in motion—flowing and transforming. Energy can change form as it proceeds in its eternal rush, but it can never cease to move.

It can also be said that the movement of energy can be seen as being purposeful and not random. The flow of energy can be understood through the observation of its movements. Clearly, energy is the essence of biology—the purposeful flow of energy allows for life forms to exist. The flow of energy can be characterized as the flow of life.

This meaningful utilization of energy is what can be referred to as the macrocreation force. This force informs life, and it formulates existence. The fact that energy has purposefulness at its core means that existence also has purposefulness at its core. A life form is not a random act; it is a determined act. This act can be sustained for eons. A life form owes its unique existence to the determined designs of energy—the macrocreation force.

Perhaps there was a stage in the development of the universe when energy did flow in a maelstrom of randomness that is counter to what can be observed today. This could be considered a "primordial" era—a time when existence consisted of nondirected bursts of energy that resulted in nonsustainable forms of "life." But, at some point, it is reasonable to conclude that this random activity began to take on a repetitive character. It is this quality of repetitiveness that allowed energy to accomplish the most rudimentary aspects of design. From this very fundamental level, increasing complexity of design developed.

To state this second principle quite succinctly, the primordial stage of nondirected energy eventually led to the growing capacity of energy to achieve repetitive designs of increasing complexity. It is the essential nature of the macrocreation force to be in ceaseless evolution, and this evolution must continue to lead to the increasing complexity of form and design. The universe that we inhabit can most certainly be characterized as quite complex in all its aspects, yet it is presumably not as infinitely complex as it shall someday be.

The Third Principle

The purposeful movement of energy that supplanted the primordial randomness of energy occurred due to one underlying factor—incipient awareness. In order for energy to engage in design, it was necessary that there be awareness of design. The flow of energy became purposeful for one reason; the purpose was provided by sparks of awareness. It was possible for energy to form the rudiments of life because energy had also acquired the rudiments of consciousness.

As the complexities of the design of energy increased, so did the associated capacity of awareness increase. Thus, it can be seen that the infinite elaborateness of the energy matrices that form the universe indicate the infinite aspects of consciousness that are intrinsic to this universe. To behold the artfully sophisticated designs of energy that this universe has to offer is to acquire evidence of the artfully sophisticated nature of awareness.

The macrocreation force—the purposeful flow of energy in all its permutations—can be considered insatiable. There can be no aspect of creation that is so glorious that it would serve to blunt the creative urge of awareness to evolve beyond it. The supreme complexity of this universe exists because the macrocreation force has brought it to this level. The marvels of creation that surround us now will be supplanted by the marvels that are to come.

There is a term that can be used to describe this form of awareness that accompanies energy—the macroconscious. The macroconscious is the awareness that is the automatic result of the designs of the macrocreation force. There will be people who will want to characterize the macroconscious as being something that is consistent with their personal concept of God. While this idea is not appealing to me on an intellectual level, there can be no final word on this issue. My view is that human beings invented "God" for both cultural and psychological reasons. As far as I am concerned, God is a mythical figure no different than Zeus.

On a rational basis, I believe that it cannot be valid to characterize the macroconscious as a glorified human being. The macroconscious is no more a human being than it is an asteroid belt. Human beings are not the epicenter of all existence and the creative force behind existence should not be seen as human. There will be much more about human vainglory and the nature of the macroconscious in the pages ahead.

A Corollary Principle—Macrotime

Because energy cannot be created or destroyed, this must mean that energy has always existed. Accordingly, this must also mean that there has always been existence. Further, all declarations that mainstream science has to offer in regards to the "age" of the universe should be considered highly dubious.

The central problem is that human beings perceive time in a linear manner. Most scientists have not made the mental effort to throw off this limiting factor. All hypotheses regarding the "age" of the universe spring from this linear mental framework. The use of a system of linear time can only lead to a conundrum of logic.

In the beginning, there was no beginning. Energy is existence, and so there has always been existence. There is no such thing as a point of origin for the universe. Regardless of human mental processes, logic demands that the actual nature of time be understood as nonlinear. A term that can be used for this nonlinear nature of time is macrotime. The reasons why humans and other biological forms of life perceive time in a linear fashion will become clear as the full scope of the macrocreation philosophy is discussed.

Because we operate in a linear mental framework, it is not easy to make the leap of logic that takes us to a nonlinear system of time. However, one way to see macrotime is in the context of three dimensions.

In this three-dimensional framework, there is no up and down, in and out, or back and forth. With this structure, time can be described as an infinite array of points, with each point representing a "point of present." Each of these points of present has an equal value with every other point. No point has a relationship of "past" or "future" to any other point. Time simply "inhabits" one point of present after another.

With this understanding of macrotime, it can be seen that the universe truly has no "age." The universe never began, and it will never end. The macroconscious operates from this system of macrotime, but human beings do not. It is a great mental challenge for people to see beyond the limitations of their linear perceptions, and even many of the scientific minds of our day have not been up to this challenge. But the true nature of this universe simply cannot be understood any other way.

Existence versus Nonexistence

The idea of existence versus nonexistence is a question that was actually settled "long ago." Logically speaking, the possibility that there might have been nonexistence rather than existence has to be considered a valid option. The manner in which existence "triumphed" over nonexistence can be looked at by using a basic analogy.

When a coin is flipped, it will land either on heads or tails—excluding the unlikely scenario in which the coin lands on its side. When the coin is flipped and it lands on heads, this quite obviously means that it did not land on tails. The only explanation that can be given for the fact that the coin landed on heads is that it did so because it did not land on tails. It had to be one or the other.

Using a truly "cosmic" coin, existence can be likened to heads and nonexistence likened to tails. Thus, we have existence because this cosmic coin was flipped and it landed on heads. Presumably, it could have landed on tails instead. Because tails is a perfectly valid alternative to heads, nonexistence has to be seen as a perfectly valid alternative to existence. Since we clearly do have existence, the issue must be considered settled.

Atheists contend that, upon death of the "physical" body, we will cease to exist, but this is not a rational conclusion. All life forms are composed of energy, and this energy cannot be destroyed. Because we have existence, we can never have nonexistence. Even if some of us may have a deep yearning for nonexistence, it is not an option. There is certainly plenty of room for debate regarding the nature of our existence upon "physical" death, but there can be no question that existence, of some type, must continue. We cannot travel down the existential road of "nothingness" because that highway dead-ended before we ever had the chance to get on it.

Some people might wonder "who" flipped the cosmic coin that decided the issue of existence versus nonexistence, but this is not a useful question. Understandably, human beings think in human-centric terms—the "who"—when it comes to these grand cosmic questions, but it should be understood that there are no people operating at the core level of existence. The flip of the coin of existence occurred in the absence of any human, or godly, involvement.

Section Three—The Macrocreation Philosophy

Opening Remarks

This section speculatively applies the trinity of principles that comprise the macrocreation theory to many aspects of existence. The ideas that are found in this section serve only as my personal and subjective analysis regarding how the trinity of principles relates to a large number of questions and topics. First and foremost, these ideas are presented as a demonstration of how the trinity of principles can be applied to a variety of issues. These ideas are not doctrine or dogma.

The macrocreation philosophy is a living instrument. It is a belief system that can incorporate the individual analysis of all those who consider the trinity of principles to be based on sound logic. Ultimately, this means that anyone who wishes to take the time and to make the effort is in a position to write a work of macrocreation philosophy that is consistent with their own particular inclinations. A person may call himself or herself a "macroist" while choosing to reject much of the material that is in this section of the book. A macroist always thinks things through for himself or herself.

The only doctrine of the macrocreation philosophy is the trinity of principles. Everything else in the belief system represents individual efforts to apply those principles to this world and to our lives. Any human being who values rational thinking is fully qualified to make these efforts. In a philosophical system that lives and breathes as it should, no one's personal analysis should ever be considered the final word.

The speculation that I offer in the pages ahead can be thought of as a springboard. Obviously, the philosophical conclusions that I present in this section are ones that have personal appeal to me. Without doubt, not all of these conclusions will appeal to everyone. Those people who find themselves in disagreement with me can then think things through for themselves and reach conclusions that make better sense to them.

The macrocreation philosophy is designed to live and develop in the mind of each person who has chosen to look at this world in this new way. In exchange for giving up faith in the unknown, the macroist embraces rational analysis of all that is known and all that is logical. There is no sacred text. There is no one to worship. There is no elite. We are on our own.

Human Vainglory

In genetic terms, humans are virtually identical to other mammals, especially the primates. Theologians contend that we humans dwell on an entirely different metaphysical plateau than do animals. But this contention certainly has no basis in biology and should be considered an example of human vainglory.

Genetics provide the information that humans and animals are significantly comparable forms of life. Another way to look at it is to say that humans and animals are on much the same path in life. No Christ figure died on the cross in order to provide salvation for chimpanzees, and this did not occur for human beings, either. Quite clearly, aardvarks, armadillos, and anteaters are not in need of salvation, and neither are any other animals, including humans. Only humans consider themselves to be God's "chosen" group of animals.

Typical New Age doctrine can also be examined in this light. Presumably, a prairie dog is not doing certain things out there on the prairie in order to rid itself of the weight of "bad karma" from a previous lifetime. It is unlikely that an antelope can be found who is on a spiritual quest for redemption or enlightenment. If any of the "spiritual" pursuits that humans may pursue are applied to the activities of any animal, the illogical nature of these pursuits can be clearly understood. New Age gurus are very much in close alignment with tradition-encrusted religions on this major issue—that human beings exist on an entirely different plateau of existence than all other forms of life, and thus humans must pursue "spiritual" goals that pertain solely to our species. But there is no logical reason why this should be the case.

No other animals but we humans would ever be guilty of the vast grandiosity with which humans are afflicted. Virtually all of humanity's religions, denominations, sects, cults, cultural groups, and various New Age philosophies focus their particular doctrines to this human-centered delusion of grandeur.

Of course, there can be no doubt that we humans do possess some unique and significant characteristics that the other animals do not share. Yet humans have elevated themselves to a position of mythological glory that must be considered irrational. This rampant self-aggrandizement has greatly impeded human attempts to achieve a rational and genuine enlightenment. The accomplishments of human culture need not be denigrated simply because humans are not, in fact, "above" the animal kingdom. Our accomplishments remain quite extraordinary in their realization.

There are certain religions and New Age philosophies that propose the concept that human beings are on a quest for some measure of perfection—this would mean the end of all "negative" thoughts and deeds. Clearly, however, this idea of perfection cannot be applied to any other form of life, such as a gorilla or a fern. Unquestionably, a gorilla and a fern are already perfect for their roles in this world—and so are human beings. A human need not be concerned with the eradication of "negative" thoughts any more than a gorilla is. The human-centered concept of spiritual perfection is nothing more than a fantasy created by humans to honor themselves.

The macrocreation philosophy contends that a human being is no more a "child of God" than a virus is. There is no Spirit in the Sky or Mother Goddess to look down upon us and provide us with approval or disapproval for our actions. Despite our fervent desire for self-glorification, we humans do not dwell in isolated splendor, separate and apart from all other aspects of life.

Every aspect of existence results from the creative designs of the macroconscious, and all forms of life respond in equal measure to the thrust of the macrocreation force. Humans are indeed unique, but not quite as steeped in glory as we would like to believe.

While there is no such thing as "spiritual growth" as it is defined by religious and New Age groups, there can most certainly be the achievement of creative growth during the course of a person's lifetime. Anyone who accomplishes a degree of mastery over his impulses and inadequacies has achieved something that is genuine and valuable. However, this creative mastery has nothing to do with the concept of getting "closer to God." An individual can grow warmer and wiser by making the right choices in life, but this is a function of creative energy and not of spirituality.

The concept of spiritual growth is something that has served to give human beings worthwhile goals to strive for, and so this has been a useful device in our society. In terms of energy, however, we are already as "perfect" as we are capable of being. We humans are an intrinsic aspect of the continuum of energy that comprises existence itself. Logically speaking, energy cannot be described as "spiritual" in nature. When we human beings come to understand our energy nature, then we can begin to redefine our goals as we learn to redefine ourselves.

The Macroconscious and the Divine

Consciousness is inextricably bound with energy. Where energy flows, consciousness flows. What can be termed the macroconscious is the totality of consciousness. The term macroconscious should not be considered synonymous with God because the traditional concept of God is essentially a human-centered notion. The macroconscious is certainly not a mythological Zeus-like figure. At its core, the nature of the macroconscious should be considered an abstraction that does not relate to anything human.

All disciplines of science show us that life has coherent organization. The molecular dynamics of any form of life display a highly sophisticated manner of organized energy. It naturally follows that consciousness itself has a highly sophisticated organization. Mere randomness of consciousness would not allow life to exist. Thus, the macroconscious can be seen as the "organism" of awareness. All aspects of existence are components of this organized awareness.

In contrast to an utterly vague term such as "spirit," the macroconscious can be understood on an analytical level. While the full nature of the macroconscious is no doubt beyond human comprehension, the macrocreation philosophy will offer ideas on how humans fit in with the greater scheme of things.

When considering the nature of the macroconscious, it is quite desirable to avoid reaching simplistic conclusions that do not allow for the infinite complexities that are involved. Nonetheless, it is worthwhile to initiate a plan of rational deductions regarding the macroconscious so that some philosophical progress can be made. Although the term "divine" is very imprecise, it is a useful term because most people have a general idea of what the "divine" relates to. People typically consider the "divine" to be a power, spirit, or god that dwells outside of us and connects us with the greater part of existence. Perhaps every single person who has ever lived has felt this connection.

The nature of this divine thing can be easily explained if the person who experiences it has strong spiritual or religious faith. However, if the person who experiences this divine connection is an atheist, agnostic, or devout rationalist, then a satisfying explanation has been much harder to come by.

Many people like to think of the divine as being the manifestation of "pure love," but there is really no rational basis for such a conclusion. The macroconscious cannot be characterized as manifesting pure love, or anything like it. Logically speaking, the macroconscious should be thought of as manifesting all aspects of life, including hatred, anger, and lust, as well as such things as compassion and harmony. It seems reasonable to say that harshness and violence are intrinsic aspects of this world we live in and are, therefore, intrinsic aspects of the macroconscious itself.

The time will surely come in the evolution of human understanding when the myths and the fairytales that have nurtured human culture for millennia will finally be put aside in favor of a more "adult" comprehension of existence. We know that fairytales are told to children in order to give them both comfort and entertainment. Myths that appeal to the child inside of us have become entrenched in the belief systems that human beings live by. Someday, as a species, we will leave childhood behind and achieve a truly adult comprehension of the complex world around us.

The Macroself

In terms of the continuum of energy, every person has an intimate link to the macroconscious. The matrix of energy organization that serves as this connection can be termed the macroself. The concept of a "greater self" that lies outside of our usual human consciousness is an idea that can be found in numerous New Age texts. Most definitely, the conceptualization of a greater self is not an original idea that I claim for myself. However, the focus of the macrocreation philosophy is to comprehend all aspects of existence in terms of energy and, thus, the macroself will be explored as it relates to the organization of energy.

Whenever a person turns to "God" or "Spirit" for solace and guidance, it is the macroself that responds. While the macroself lies between the human self and the macroconscious, the macroself is more aligned with the motivational force of the macroconscious rather than the human self. This means that the macroself's perspective is far removed from the perspective of the human self. The macroself does not act in a manner designed to guarantee comfort, happiness, and satisfaction to a human being. In fact, the macroself may lead the human self into discomfort, unhappiness, and utter dissatisfaction if this is what is in accordance with the macroconscious. It should be clearly understood that the macroself is no "guardian angel."

As discussed previously, the macrocreation force is not something that operates from a motivation of typical human-centered benevolence. The idea that the fundamental force of existence can be characterized as something that is comparable to "pure love" is highly naive. The force that underlies existence is responsible for all aspects of life, including those things that we bemoan and abhor. The macroself is in sync with the macrocreation force and will behave accordingly, regardless of human desires.

Even though the macroself is capable of leading the human self into traumas and troubles, it should be understood that the motivation of the macroself is to assist us in living the most creatively valid life that is available to us. It should be very clear to one and all that this earth reality that we all inhabit is not a paradise. This world is a place where turbulence and tragedies routinely occur. It is the contention of the macrocreation philosophy that these harsh aspects of life are intrinsic to this world by means of creative design.

There are those who choose to believe that our current reality represents a paradise that has been lost. I believe that, due to the innate nature of the macrocreation force, paradise is not a possibility. The macroself maneuvers the human self into a lifetime of experiences that fulfills a necessary creative function. Although the pain and torment that we endure in this world is a necessity, sufficient relief from this suffering is also provided, as it is needed. Our lives play out in a dramatic reality that represents the dynamic thrust of the macrocreation force. As Shakespeare once stated, all the world **is** a stage.

However, it needs to be stressed that the human self is not the helpless puppet of the macroself. The macroself can be described as a focused matrix of individualized energy and so can the human self. While there is no doubt that the macroself has a more powerful store of energy at its disposal, the human self should never be thought of as powerless.

The macroself provides creative direction to the human self, but it cannot **command** the human self, as a general rule, to do its bidding. The macroself will provide an extremely strong intuitive direction for the human self to take, but the human self does have the ability to resist this direction. The interplay of creative energy between the macroself and the human self should be understood as being utterly complex and subtle. In no way is it simplistic and one-dimensional.

For all of human history, almost everyone has realized that they have a genuine connection to the world that exists around them. Most of us understand that we do not exist as wholly isolated entities but have a link to other people, animals, nature, and the entirety of our encapsulating realm. What has not been understood is that this link is due to the continuum of energy that we are all a part of.

Theologies have provided highly imaginative scenarios that attempt to explain this sense of connection that we have with all living things. Billions of people have adopted these explanations because they were appealing in some way and because they were the most ready explanations available. In contrast, scientific atheists proposed the idea that this sense of connection was mere imagination, because the nature of the religious explanations was so imaginative and quite implausible.

These differing points of view created a philosophical stalemate that has endured for generations.

It is the intent of the macrocreation philosophy to offer a coherent explanation for the connection that people feel for the world that exists around them that does not rely on myths, fairytales, or irrational theologies. The rational contention can be made that consciousness, like all aspects of life, is organized in a fully logical manner. The macroself is simply part of that logical organization.

On a purely rational basis, the theologies that form part of our cultural landscape today should have become marginalized in our society in the same way that the myths of various ancient cultures are thought of as being merely quaint and charming today. However, what has allowed these abundantly irrational theologies to survive and play a major role in our world today can be characterized as their rich "soul-satisfying" appeal. The power of this appeal is undeniable. In sharp contrast, the macrocreation belief system has little, if any, ability to inspire and provoke moments of sheer rapture.

The concept of a personal God who loves us and watches over us at all times is an understandably appealing one. It especially appeals to the child inside all of us. It is necessary to put aside these simple, but heartfelt, emotions if we are to see the world in a truly clear light. It must also be considered a rational option, however, to choose not to see this world in a clear light. The naked truth can be an unappealing truth.

In order to accomplish its creative maneuverings, the macroself will work with the belief systems that the human self employs. This is the case even if these maneuverings serve to strengthen the religious or other belief systems of the human self. Generally speaking, it would serve no useful or valid purpose to destroy a person's belief structure. Thus, it can happen that the workings of the macroself may intentionally lead to the strengthening of an individual's religious beliefs, with even more devotion being given to it. What matters most to the macroself is creative direction, not the fostering of "truth" in its dealings with the human self.

For example, a Christian who is seeking guidance from "God" may open the Bible and pray that a particular passage may "jump out" at them and provide them with the direction that they need. The macroself is perfectly capable of responding to this situation. It will initiate a creative maneuver that allows for the Bible to provide the thing that is being asked for. Of course, the macroself can also work the same way with any religious text or with such devices as tea leaves or the I Ching. If the human self is truly in need of guidance at that time, then guidance will be provided.

Living our lives in this dramatic reality is immensely challenging, and the macroself is able to assist us in coping with our travails. As a general rule, the human self can trust the guidance that is intuitively received from the macroself. It should never be forgotten, though, that upon occasion, the macroself may lead us into trouble and not away from it. For every one of us, the difficulties that we experience are creative necessities.

Although the macroself will always have valid creative reasons for the maneuverings that it undertakes, occasionally these maneuverings may lead to the experiencing of a level of pain and suffering that the human self may consider to be unacceptably severe. In these instances, the human self can assert its own individual power and thus resist the full brunt of the pain. There will be more about this topic in the chapter that deals with illness and injury.

From the perspective of the macroself, a human lifetime that is awash in intense drama is far more creatively rich and satisfying than a human lifetime that is bland and quite uneventful. The greater the challenges that the human self is subject to, the greater the ultimate satisfaction the macroself will experience. So, it can be seen that its motivation is very clear.

While we humans may frequently desire nothing more than an endless series of quiet nights spent at home, the macroself will often have something more dramatic in mind. It is the creative impetus of the macroself to design intense, emotional adventures for us to experience, and this creative force will generally overrule the natural human desire for tranquility and easy times. Earth is a dramatic reality, and this drama will, of necessity, manifest itself with the assistance of the macroself.

Prayer

A prayer that is made to God or to a saint or to an angel or to anything else is actually a communication that goes to the macroself. It does not make any difference at all what is in the mind, or the heart, of the supplicant. Because an individual thinks that he is praying to someone in particular does not have any bearing on who or what is in a position to respond to that prayer. Only the macroself can receive prayers, and only the macroself can act upon those prayers.

As is the case with the mythical figure of "God," all prayers are heard by the macroself but not all prayers are answered. Only those prayers that are in line with the overall creative direction set by the macroself are likely to receive a positive response. The macroself has as its primary motivation the macrocreation force, and this is not necessarily a wholly benevolent entity. Frivolous and foolish prayers will be shot down without sympathy.

In many religious traditions, prayer is something that is approached in the same manner as a peon would beg the lord of the castle for a personal favor. The supplicant may go down on bended knee, with their eyes humbly downcast, and make their requests known in a hushed, respectful voice. They will implore their lord God to please grant this simple request from their humble servant. They will engage in this behavior in an attempt to get on God's "good side" and thus help ensure that their request might be granted. To approach prayer in this manner is to confirm the status of God as nothing more than a human being with a lot of power.

In contrast, the macroself requires no worship or any deferential treatment of any kind. The macroself is best understood as a matrix of individualized energy—it is not a "supernatural" being. The macroself cannot be seen as the "spiritual" parent of the human self. In essence, the macroself is that aspect of the continuum of energy that connects human energy with the nonhuman energy that forms the great bulk of macroreality.

A prayer to the macroself should be made when a person feels the need to communicate in an overt manner with the macroself. Although the macroself and the human self are always intimately linked and every one of our thoughts is "heard" by the macroself, a prayer serves to provide extra emphasis and a sense of urgency to the ongoing communication process. A prayer is like a shout, or even a scream, and the macroself will take this heartfelt plea into consideration, so long as it can serve a useful creative function.

In the case of group prayer, such as a prayer circle praying for "world peace," these pleas will be collectively heard at the general macroself level of consciousness. The response to such group prayer efforts will be in line with the broad creative directions instituted by the macroconscious. Therefore, getting a group of people together, even millions of them, with the idea that God will be so impressed that he will immediately bring peace to a war-torn world could not be more simplistically naive.

The macroself and the macroconscious do not respond to the good intentions of human beings. Utterly sincere pleas and begging by humanity will have little effect on those vastly complex matrices of energy that cannot be considered human in any fundamental way. We do not have a Father in heaven, no matter how much we might wish that we did.

Nonetheless, the macroself is capable of providing that calming sense of peace that many people do truly experience as part of their prayer efforts. If the human self is very much in need of this tranquilizing relief, the macroself will provide it as a practical means of helping us cope with the frequently harsh challenges of this world. Although the macroself can in no way be characterized as manifesting "pure love," it is capable of acting as if it were.

The Human Drama

It can be assumed that this earth reality fulfills an essential creative purpose that cannot be fulfilled in any other way. This reality can be thought of as a unique art form that is not duplicated in all the vastness of what can be termed macroreality. Earth is a microreality where the rigidity of "physical" concreteness serves as the artistic means of expression. We dwell here because the macrocreation force makes it necessary that we dwell here.

In this dramatic reality, that which is immortal exists as mortal; that which is unlimited exists as limited; the infinite experiences finiteness; and the all-knowing feels what it is like to be unknowing. The wordless abstractions of macroreality take on the qualities of concrete verbosity. Earth is art.

The macrocreation force manifests itself in every aspect of this world. It is why the writer writes, the painter paints, the composer composes, and the architect designs. It is also why lovers love and burglars burgle. The force of macrocreation can be seen in the **doing** of all things by all forms of life. Every single act can be understood as a creative act. Every activity of life can be seen as an expression of art.

In this dramatic earth reality, the macroconscious has the opportunity to experience creative challenges and rewards that would not be possible otherwise. This arena of faux physicality offers a rich array of limitations, handicaps, frustrations, hardships, restrictions, and challenges of every possible kind. We who dwell here are testing ourselves in the harshest of conditions in order that we may experience the greatest of satisfactions. We are all Olympians.

Earth reality can be likened to a fictive reality, such as what is found in a novel or stage play or movie. Each one of us is the central player in a saga that unfolds moment by moment, day by day, and year by year. We also perform important supporting roles in other people's dramas. Not infrequently, we also serve as bit players and extras as well. In our lifetimes, our roles will run the full range of possibilities. We will go from the bloom of youth to the decrepitude of old age. The diversity of our experiences will generally be vast and fascinating.

Everything about us is an aspect of creative design, from the way we look to the way we walk to the way we talk and think and feel. We exist as a matrix of individualized energy patterns. Every act that we commit in our lives will determine how those energy patterns will evolve. Our personal evolution is the result of our personal creativity.

The challenges that we face in our lifetimes—such as obtaining love and support, acquiring food and shelter, achieving fulfillment and relaxation—are actually creative challenges of the most basic kind. Because this reality usually feels like the **only** reality to us, we embrace the frustrations and the hardships rather than resisting them. However, at a deep instinctive level, we do have the sense that a valid purpose of some sort is being fulfilled as we labor through the aggravations of our daily existence, and this is what enables us to endure.

Pain and heartbreak are intrinsic aspects of this dramatic reality, and sometimes, these harsh facets of life can push us to our limits. Any popular novel or movie can be seen as an example of "art imitating art." *Gone With the Wind* is a good example of this. Toward the end of the narrative, the author, Margaret Mitchell, chooses to have the charming little daughter of Rhett and Scarlett die in a freakish accident. As the sole author of the story, Ms. Mitchell could have taken the plot in any direction of her choosing, so the question can be asked—Why did she make this harsh choice?

The answer to this question lies at the heart of this dramatic earth reality. She made the choice that she did because she considered it to be dramatically the most meaningful choice—a choice that would involve the reader the most. That was her basic motivation. It was her desire that the story she was telling should be as vivid and memorable as she could make it. This is why she chose to have the little girl die.

Every single day, we see stories in the news that mirror the drama, heartbreak, and tragedy that we can find in any piece of compelling fiction. The "rules" that drive the creation of vivid works of fiction also drive the events of this fictive reality. A little girl does not wander into a street and get run over because God was in a child-killing mood that day. Each death, each life, each and every act serves a valid creative purpose. The fact that we may suffer greatly as an accompaniment to some of these creative acts is considered an acceptable corollary by the macroconscious.

The lives that we live here are designed to be dramatic, painful, compelling, and meaningful. Consider the tedium of reading a biography of an individual who lived a long and entirely uneventful life. Such a book would be so boring it would be utterly unreadable. The most interesting books that we can read are those biographies of people who have lived lives filled with traumas and troubles and difficult challenges that had to be met. These are the lives that we enjoy reading about because these are stories that are the most dramatically compelling. This is true with our lives as well.

Every life that is lived in this earth reality is a story that is being told. If the pain and the heartbreak that we experience in our lives cannot be appreciated now, it will be appreciated later, once the story is concluded. The more dramatic the life has been, the greater the creative satisfaction that will ultimately be experienced. This fact does not necessarily make the pain more endurable, but it is helpful to understand the highly dramatic nature of our lives.

Someday, for each one of us, the stage curtain will be lowered, and we will be able to walk offstage (to generous applause, hopefully) and thus enter the "real" world that is the macroreality that awaits us. From that point on, we need perform no more.

The Abstract Core

The macrocreation philosophy proposes the idea that our earth reality represents a "concrete extreme" of existence. This means that we live in a microreality that contains parameters of presumed rigidity and immutability. The laws that relate to this "physical" world are presumed to dictate the way that things absolutely must work. In this realm, both "physical" and psychological concreteness are the most dominating aspects of our framework of reality. But this manner of concreteness represents an extreme representation of reality, not a moderate one.

Philosophical frameworks that human beings devise that arise entirely from this concrete extreme are most certainly going to be distorted and human-centered. A perspective that is completely limited to the confines of false notions of physicality is a perspective that will have little overall validity. To put it another way, to understand humanity properly, it is necessary that a philosopher escape the strict limitations of the human point of view.

The macrocreation philosophy attempts to offer ideas that arise from a perspective that is, to some degree, nonhuman. This macroperspective is not limited to the usual humanistic nature of previous philosophies but extends far beyond it, to the much broader considerations that are found in macroreality. The typical human viewpoint should be seen as an extreme viewpoint, because it relates to the extreme concrete nature of earth reality. Genuine objectivity can only be achieved by going beyond this highly limited human philosophical cage.

The full nature of macroreality, to the extent that it can be deduced, will be discussed elsewhere. What will be discussed in this chapter is what lies at the other extreme level of existence—the place farthest removed from the concrete nature of earth reality. This place can be termed the abstract core.

In our reality, what is experienced is the ultimate manifestation of **being**—whether this is human being or any other manner of being. In contrast, at the abstract core, there is an infinitesimal sense of being, something that, by our concrete standards, approximates nonexistence. The abstract core is the antithesis of all things human and all things earthly.

The abstract core is the cold, still heart of existence and underlies all aspects of existence, including our own. Earth reality can be thought of as a constructed layer of reality that floats far above the deep-seated foundation of the abstract. We see an indication of this abstractness in our nightly dreams—those symbols and images that do not seem to make any sense yet possess a potency of meaning that cannot be easily dismissed. In the dream reality, we blend the concrete characteristics of this earth reality with the infinitely mutable nature of macroreality. Our dreams can even take us all the way to the abstract core itself.

Human beings are the fullest embodiment of concreteness, and thus, it is our tendency to venerate it and to do all that we can to uphold it. This is quite understandable, but unquestioned concrete thinking is a philosophical steel trap that confines an individual to a greatly distorted point of view. All manner of fundamentalist thinking is the clear representation of this concrete constriction, and it has afflicted society to a very unhealthy degree. In order to escape from the delusions of fundamentalism and other forms of concrete thought, we must learn to appreciate the realm of the abstract. Without this appreciation, human philosophy will not be able to evolve as it desperately needs to.

Physical Laws

Physical laws are not really laws at all; they are, in fact, conventions of "physical" behavior. The system of laws that scientific fundamentalists hold sacred is arbitrary, unyielding, and mindlessly mechanistic. Because the force of macrocreation that underlies existence is a creative force, an utterly rigid concept of physical laws is not going to hold up as being truly rational.

To a significant extent, most human beings derive true comfort in the belief that this is a universe governed by unalterable physical behavior. Most people cherish the set of laws that promise to be highly predictable and inerrant. The greater the degree of internal anxiety that an individual is subject to, the greater the likelihood is that he or she will embrace whatever belief system offers the "rule of law."

As it turns out, the physical conventions that are at the heart of this world are nearly as predictable as the mechanistic physical laws that are supposedly in place. However, unlike these laws, physical conventions are not scientific absolutes—they are not black and white. It is the macrocreation force that provides the shades of gray.

The theory of gravity was said to be inspired by the observation of the simple act of an apple falling from a tree. Thus, the "law" of gravity was laid down. This law stated that there are no available options—if the apple becomes detached from the tree, then it must fall to the ground. There is no doubt that this is a reliable occurrence—a billion times out of a billion times, the apple does indeed fall to the ground.

However, the underlying nature of the macrocreation force requires that the law of gravity be suspended if there is sufficient creative cause to overrule it in a particular incident. This suspension of a physical law may take place once in a thousand years, or it may take place quite often, but the frequency of this suspension is irrelevant. If there is a really good reason for the apple not to fall from the tree once every millennium or so, then this is what will occur.

Physical laws are not absolutes; they are conventions. Only the macrocreation force is absolute. We do not dwell in a hopelessly mechanistic universe but rather in a richly creative one. Religious fundamentalists cherish their rigid laws of God, and scientific fundamentalists cherish their rigid laws of science, but both groups need to accept a world that is motivated by the creative impulse, which is capable of taking us anywhere.

The Uncommon Natural

The term "supernatural" is rather nondescriptive and can be considered obsolete, in terms of what the macrocreation philosophy has to offer. In this new way of thinking, all events, no matter what they are, should be considered events of the natural. In the past, an occurrence may have been labeled as "supernatural" based partly, at least, on the rarity of its occurrence. Frogs may fall from the sky once in a very long period of time, but this infrequency of its occurrence says nothing about the naturalness of the event.

The macrocreation philosophy offers the term "uncommon natural" as a replacement for the traditional concept of the supernatural. One way to examine the idea of the uncommon natural is to examine the occurrences of "miracles" in our world. Rationalists have tended to believe that if a so-called "miraculous" event was certified as having actually taken place, then this certification would thereby serve to validate whatever religion or belief system that was attached to that event. But this is not the case.

Scientific fundamentalists have backed themselves into a tight little corner of delusion and denial when it comes to the subject of miraculous happenings. It has been their quite understandable desire to not say anything that would validate religious ideology, but they have not examined the complete nature of inexplicable occurrences. In other words, the rationalists have not acted in a fully rational manner.

It is conceivable that every single miracle that has been attributed to Jesus, for instance, did indeed occur as reported. For creative purposes, the physical laws that would have made these miracles impossible could have been overruled. The fact that these miraculous events served the burgeoning Christian faith is presumably due to the fact that this bolstering was a desired creative device. The creative maneuverings of the macroconscious can propel one religion to the forefront in one particular era then send it crumbling into the ashes in another. The macrocreation force determines the fate of kings, and even gods.

Those religions that play a prominent role in world affairs do so purely as a result of creative design. The macrocreation force allows for the occurrence of innumerable miracles, large or small, as creative programming. These miraculous events do not serve, in the tiniest way, to validate the theology of any religion. These miracles can, however, validate the social position of a religion. A 100-foot tall image of Jesus or the Buddha appearing in the skies of a major city will certainly have a significant impact on recruiting efforts, and this would be because it fulfills an overall creative goal for human society.

Over the years, there have been numerous reports of various people seeing images of Jesus or the Virgin Mary appearing in such places as screen doors, tortillas, and just about anywhere else. There have also been many sightings of religious icons that appear to "weep" in an inexplicable manner. Scientific fundamentalists have no options when dealing with such phenomena—they can only stridently insist that none of these things can possibly be real. Intellectually speaking, scientific fundamentalists have nowhere to go. But the macrocreation philosophy shows that such odd occurrences as a weeping statue of Jesus can be perfectly valid events.

Every single religion that exists in the world today serves a useful creative purpose. Therefore, the force of macrocreation will, upon occasion, be employed to manifest an incident of the uncommon natural in order to advance the promulgation of a particular religion because this will serve the goals of an overall dramatic design.

The power that religions have to influence people's lives is very real and very potent. Nothing about this power should be considered delusional. Miracles can be quite real and quite creatively powerful. This genuine power is not negated by the fact that no theology has any measure of truth contained within it. Miracles occur to support religions, because religions exist to support people.

While some people will be thunderstruck with absolute awe by a particular miraculous event, other people do have the very rational option of ignoring completely this very same event. The image of Jesus or Vishnu appearing on a teapot in somebody's kitchen may be a very relevant thing for a great number of people, but it need not have any special relevance at all for everybody else. Even if a 100-foot tall depiction of the Last Supper appears in the skies over Las Vegas, this theatrically miraculous occasion can still be safely ignored by all those who choose to ignore it.

It is also possible that a miraculous occurrence may manifest itself simply for your benefit and have no meaning for anybody else, even members of your own family. When it comes to any incidents of the uncommon natural, it is always best to never make any broad assumptions. A miracle is never truly miraculous; it is simply a creative tool. Whenever contemplating the nature of a miraculous event, try to make an effort to understand the reality of the creative maneuverings that lie behind it. Do not be unduly impressed by any miraculous occurrence, no matter what it is.

Religious Figures in History

The potency of religions in the human heart is very much by design. Religions have played an extremely major role in human society because they possess the inherent creative power that allows them to flourish as they have.

A scientific fundamentalist would make the claim that all religions have succeeded as well as they have due primarily to human gullibility and our nearly unlimited capacity for self-delusion. Religions have also been labeled an "opiate," affecting human behavior like any other drug. These harsh opinions pertaining to religion relate primarily to the quite obviously irrational theologies that the world's religions have had to offer. But the irrationality of religious doctrine has nothing to do with the very real creative power that religions possess.

Throughout human history, religious figures—both major and minor—have served as catalysts for innumerable religious movements. These figures range from the acknowledged giants such as Abraham, Moses, Jesus, Mohammad, and the Buddha to such lesser lights as Joseph Smith, Mary Baker Eddy, and L. Ron Hubbard. Even further down the list are those leaders of small cults who have had a tremendous impact on their respective groups. All of these leaders have had at least one thing in common: the potent creative power at their disposal.

In terms of energy, it is simply not true that "All men are created equal." The amount of personal creative power that an individual possesses varies widely—there is no comparability whatsoever from one person to another. When it comes to manifestation of energy, some people can be characterized as white-hot "stars," while others can be considered as mere low-wattage bulbs. Some people are vastly wealthy in energetic terms, and many others are paupers. A very select group of individuals will possess a certain quantity and quality of creative energy that allows them to alter the course of human events.

Utilizing a poetic analogy, it can be said that Jesus, Mohammad, and the Buddha serve as the "suns" that provide the power for their respective faiths. These extraordinary human beings possessed the fire and the light that ignited their religions and inaugurated the paths that would be followed generation after generation. In their cases, the term "god" can be defined as a person who embodies the very pinnacle of personal power. They are fully human in every way, but they have the ability to affect other people in a way that can scarcely be overstated.

While there can be no doubt, from an objectively rational point of view, that the theological doctrines that form the essence of all major religions must be attributed to myth and fantasy, the power that these religions have in the general population must be acknowledged as quite real and legitimate in their own way. It is irrational to believe that the billions of religious adherents that can be found in this world have all fallen prey to the same type of opiate. Religions dominate society as they do because of the creative power that they possess; this is the most truly rational conclusion that can be drawn.

The primary purpose that religions serve is to provide both a belief system and a support system that the average person can utilize in their daily lives. Since this function is so vitally important, religions, and religious leaders, have been vested with the creative energy necessary to ensure that this function can be fulfilled. It is simply not the least bit feasible that most people should make the effort to confront the mysteries of existence on a routine basis. Religions lift that burden from people's lives by offering both a belief and a support system that is typically a part of their cultural background. These theological frameworks allow most adherents to go about the activities of their daily lives without continually pondering the nature of their existence.

Those individuals who served to inaugurate religious movements were able to do so because their personal energy matrices made it possible for them to do so. Any person who can be considered a leader in society acquired that position because of the creative energy at their disposal. This is true not only for religious leaders, but for also for many others, including military and political leaders, major writers and artists of all types, show business stars, and anyone else who has the personal magnetism to command the spirits of the multitudes.

Those important leaders who can motivate masses of people to do their bidding should not be thought of as accidents of history. A religion that holds the allegiance of hundreds of millions of people did not come about by mere happenstance. A significant figure in human history can be thought of as an instrument of energy. These instruments of energy serve to advance the creative design of the macroconscious. Our human history should be thought of as a dramatic saga that satisfies the unwavering impetus of the macrocreation force. Earth history is the art of storytelling taken to the very ultimate degree. In order for these stories to unfold as they must, there needs to be major protagonists who act in a manner that cause great creative directions to spin forth. This world can only make sense on a creative basis.

It is not too difficult to speculate on the particular creative functions that the life of Jesus fulfilled. The Old Testament epoch was an era of unbridled tyranny in which a Zeus-like God appeared to act like a nasty human despot. This was a God who was filled with imperious wrath and holy rage. These Old Testament times were not limited to the Jewish people alone but also characterized much of the "civilized" world at that time.

One conclusion that can be made is that this era of Old Testament tyranny represented the fulfillment of certain creative urges of the macroconscious. This time of terror, vengeance, and pestilence may seem quite horrific to us now, but it was presumably as creatively necessary as was the Age of the Dinosaurs. The various epochs of human history can be understood as creative developmental stages—quite essential for the sake of vital creative growth.

The life of Jesus marked the conclusion of one creative era because the manifestation of the Old Testament creative drive had been fully achieved and it was time to move on to the next. In contrast to the previous epoch, the New Testament age represents a big step toward a much more humane and enlightened framework in which to live. Jesus brought forth words and deeds that allowed for a much gentler concept of divinity. "God" was wondrously re-invented.

The "Christ" that presented himself to the world was obviously not a despotic Zeus-like personage. Jesus came along as the "Son" of God, and every person knows how much different a son can be from his father. This fact allowed people to look upon Jesus as someone who would supplant his father as a new-generation God of moderation and tolerance. In fact, the role that Jesus played as Savior was, in a very real sense, to save us all from the mindless wrath of God.

In truth, Jesus was not the "Son," but rather a sun of instrumental energy. Even though the geographical area that Jesus dwelt in was quite small, I believe that his personal power was so great that it affected the world as a whole. The matrix of individualized energy that Jesus possessed was massive enough, presumably, to have a significant impact on the totality of energy that comprises earth reality itself. This mammoth impact would also have been manifested by such figures as Mohammad and the Buddha.

Sacred texts can also be looked at as instruments of energy. It is possible for certain books to affect a person in much the same way that a charismatic religious figure can. The words and language found in venerated religious texts have a power that is genuine and not merely mythical. The devotion that an individual can show for his sacred text is not due to delusion but is a response to the actual power that the text manifests.

It should be noted that the relationship of power that exists between a person and a book is very much a personal thing. The Bible, for instance, may have a tremendous impact on millions of people but will not be able to impact the lives of millions more. This is true for all texts, of course, both religious and nonreligious.

Someone whose life has been immeasurably affected by reading and studying the Bible, or the Koran, may find it very hard to accept that this power of words has no universal impact. No book has automatic power, no matter what book it is and no matter how many people it has affected. For every person, there will be a creative interplay between any text and themselves, and this interplay is a unique phenomenon each time. For psychological reasons, people want to believe in the universal power of a sacred text, but there is no universality at all. For some, the Bible or the Koran will be a light that shines like the sun; for others, it is nothing more than a vague shadow of no real importance.

"In the beginning, there was the word." A word is a creative device that provides a concrete representation of something that is symbolic and complex. A word makes that which is an abstraction seem specifically comprehensible. A word contains the wordless—the sophisticated abstraction that lies within. The nature of language provides a glimpse into the nature of reality—a layer of concreteness laid upon the subsurface of abstraction.

The creative power that religions have manifested in human society really cannot be overstated. Religious and spiritual inspiration have greatly impacted all the human arts—drama, music, literature, architecture, fine art, and any other area that can be thought of. For the entire history of humanity, religious inspiration has been the mighty engine of creative expression. The mythic characters that the world's religions have produced will resonate with human beings for all time.

However, it appears that we live in an era where this creative power is on the wane. The unstoppable force of cultural evolution has brought us to a place where it has become increasingly difficult to take these mythological theologies as seriously as we once did. Even the greatest of creative forces do not exist outside the constraints of evolutionary creation. Thus, we appear to be on the cusp of a true New Age, an age that is defined by rationality and comprehension, and not by mythological glory and wonder.

The Proliferation of Religions

The abundance of religious belief systems in the world has a direct relationship to the abundance of the personality characteristics of human beings. For a belief system to seem valid to any individual, that system must be compatible with that person's psychological makeup.

Religions have a variety of psychological imprints that are imbedded in their doctrines and dogmas. Rather obviously, those people who possess a rigid mindset will respond most favorably to a belief system that provides black-and-white parameters for dealing with such thorny issues as morality. Clearly, someone who can be characterized as a religious fundamentalist (or a scientific fundamentalist) can also be considered a psychological fundamentalist.

At the other end of the psychological spectrum can be found such people as the Unitarians (and some others), who feel that it is somehow intellectually improper to reach any conclusions regarding anything. This psychological type believes that "asking the right questions" is much more important than finding answers to these questions—ever. Unitarians are so addicted to philosophical discourse that they are fundamentally unable to achieve results of any kind.

Most human beings fall within the two extremes that are represented by fundamentalists and Unitarians. We are all on a search for answers that make sense. For many people, it is more important that their belief system make emotional sense rather than logical sense. These two criteria seldom run hand in hand. It can be said that "theologies are mythologies that relate to psychologies." As stated earlier, it is my contention that religions have never served a role as delivery systems for philosophical truth, even if they claim that this is precisely their role.

Throughout history, religions have been able to provide humanity with moral guidelines that have proved to be somewhat useful. But, of course, moral guidance can also be included in a system of belief that has no mythological framework whatsoever. Thus far in human history, the psychological practicality of religious systems have certainly overwhelmed their rational impracticality. Our need for religions has been primarily an emotional need—it has been the neediness of an insecure child. However, adulthood may finally be on its way.

Despite the grandiose claim of the Christian church to being the "universal church," no single religion can meet the emotional needs of all the different psychological types who inhabit this earth. Someone who is a devout Buddhist has a psychological makeup suitable for Buddhism and is not a likely candidate for conversion to Islam or some other rigid, Western-tradition religion. Hundreds of millions of people—Buddhists, Hindus, pagans, Wiccans, and many other groups—are, by their fundamental nature, utterly incompatible with the religious systems of the Western tradition. Like oil and water, certain psycho-social combinations are simply impossible.

As a general rule, the religious tradition in a society that is most dominant will be the one that feels the most compelling to most members of that society. As a rule, with clear exceptions, the religion that provides the greatest social support will be the religion that feels the most valid to someone in search of a belief system. This is certainly not surprising. Few people have the strong desire to conduct a truly exhaustive quest for the "best possible" belief system out of the multitude that is available. Since people require a support system just as much as a belief system, the most convenient course of action is simply to combine the two.

In contrast to this majority position, some people will reject the mainstream religions of their particular society simply because their individualistic tendencies will lead them in a different direction. These individuals will be attracted to "fringe" systems that appeal to their basic personality types. The theologies that various people find compelling really have much more to do with psychological considerations rather than any intellectual ones. Those who consider themselves the ultimate rebels may opt for some manner of Satanism for the outcast image it has to offer. For every personality type, there will be a religion or other philosophical system that will satisfy them.

All things considered, humanity has never actually been engaged in a serious "search for truth." Instead, human beings have been on a never-ending quest for psycho-cultural satisfaction. This all-consuming quest has far outweighed the theoretical concept of "truth," even though "truth" is what all the religions of the world have claimed to provide.

The Occult Arts

On the basis of surface appearance, a rationalist will not unreasonably reject the possibility that any occult art—such as astrology, numerology, palmistry, voodoo, tea leaves, and many others—could have any degree of validity. Yet the macrocreation philosophy demonstrates how it can be possible for the occult arts to be both absurd and valid as well.

The macrocreation force is the underlying force for the action of all things. Since this is a creative force, it is not necessary that all actions that take place as a result of this force be logical or sensible. Taking the example of astrology, for instance, the proposition that the movement of the stars has a significant impact on our lives could not be more irrational, yet the macroconscious can make use of this bizarre system for its own creative purposes. That which is invalid on a rational basis can be made valid on a creative basis.

If a particular individual has allowed astrology or some other occult art to be a part of his or her life, then that person's macroself can utilize that fact for its creative maneuverings. As is the case with a religious system, the macroself may choose to act in a manner that may serve to reinforce an occult system in order to accomplish certain creative goals.

Many of the practitioners of the occult arts are quite intelligent and reasonable people. These people utilize the occult arts because at the macroself level, this use has been supported and nurtured. For example, if a person's astrological chart indicates that a particular date will be a day of "unexpected travel," the macroself, for its own purposes, can make this happen.

However, it can be seen that the macroself does not ever provide the human self with the sense that an occult system is foolproof. As anyone who has practiced the occult arts knows, sometimes the crystal ball is clear, and sometimes it is hazy—both literally and figuratively. While there will always be a few people who go overboard in their use of the occult arts, for most people, the obvious imperfections of any occult system helps to keep them from becoming too dependent.

Many occult arts involve a significant amount of intuition on the part of the practitioner. This is certainly true in such activities as reading tea leaves, interpreting tarot cards, and studying astrological charts. This intuitive insight comes from the practitioner's link with the macroself. For its part, the macroself will have the requisite quality of objective perspective that will serve to usefully guide the human self.

While it can be said that the numerous occult arts have no innate validity and can only function by means of the macrocreation force, this is true for all disciplines of science as well. Physics, biology, and astronomy all have the surface validity and appearance of rationality that the occult arts lack, yet they all serve the purposes of the macrocreation force. The only validity that any discipline of science has is, ultimately, creative validity and not an arbitrary, mechanistic validity.

It is always a mistake for the practitioner of an occult art to reach simplistic conclusions as to the functioning of that occult system. For every instance where an occult practice may prove helpful, there may be a corresponding instance where it does not prove helpful. No occult art is fully reliable because the creative maneuverings of the macroself are simply too unpredictable.

One highly simplistic concept is the idea of affirmations. There are many people who actually believe that making the statement, "I will become wealthy," will automatically lead to this occurring. Reality does not work this way. The acquisition of great wealth will have to be in line with the creative goals of the macroself in order for this to happen. Wishing will not make it so.

It is deeply ingrained in human nature to want to believe that our lives are governed by mechanistic systems, whether these systems be occult or science-based. But no one should ever put complete faith in any system, because any system can be overridden by the macrocreation force. A world that lacks **any** truly arbitrary rules is a scary concept for most people, but the creative nature of our existence means that we are never at the complete mercy of a mechanistic, rigid, fundamentalist set of laws.

The Physical Universe and Alien Life

The idea that we all live in an infinite physical universe has been accepted scientific dogma for many generations. The number of galaxies that supposedly spread out forever is said to be incalculable. Despite the fact that physical infinity is completely illogical, this concept has not been contested in spirited debate.

In contrast, the concept of an infinite nonphysical universe is wholly logical. To begin with, energy is clearly infinite—no limitations can be placed upon it. When it comes to matter, however, the amount of matter, whether it be ordinary or "dark," needed to construct a limitless array of galaxies is truly mind-boggling. The whole question of "dark" matter really creates more intellectual problems than it solves.

An infinite concrete universe is a conundrum, while a universe that owes its existence to being a presentation of energy is not a conundrum. Because no limitations can be placed upon energy, there is no such thing as finiteness. The universe that we perceive is a display of concrete form, not concrete form itself. The reality that we perceive as existing "out there" is a designed reality, and we perceive it within the limitations of our own designed reality.

It is the contention of the macrocreation philosophy that all that we see in the night sky is actually a vast, dazzling backdrop to our earthly affairs. Without doubt, it is a universe that has the capability to fascinate us with its mysterious complexities, but these aspects of intrigue serve to provide us with a "sound-and-light show" that is there to enhance the dramatic reality of earth. Quite understandably, modern scientists have focused their attention on these fantastic energy presentations in much the same way that an infant will focus its eyes on the colorful string of plastic baubles that hang above its crib.

Another aspect of scientific thought that has become dogma to most is that this universe came into existence as a result of a Big Bang and that we are still feeling the effects of this Big Bang even today. However, even the Big Bang, assuming such a thing ever happened, should not be considered the causal factor for the manifestation of this universe. The only true causal factor was the creative impetus that was provided by the macrocreation force. All other aspects of existence, including the Big Bang itself, can be thought of as incidental.

If earth is looked at as a microreality of some sort, then some consideration can be given to the boundaries of this microreality— where does the microreality end and the macroreality begin? Perhaps an answer to this question will come from future expeditions of long-distance space travel. If, for instance, astronauts who pass a certain point in their journey—possibly the edge of the solar system—begin to experience an onslaught of hallucinatory images, then this may indicate that the outer boundaries of this earth-based microreality have been reached. This presumably would be a boundary that could not be breached.

Another possibility is that space travelers would be able to take their earth microreality with them, like a bubble, and thus there would be no limits as to how far they could go. We simply have not gone far enough into the deeper reaches of space to make any determination as to our possible limits.

If our universe can be thought of as being infinite in any way, it is because our imagination is infinite. The concept that the universe is composed of an infinity of physical matter defies a genuine application of the principles of logic. However, the idea of nonphysical infinity, such as what can be experienced in the dream state, is quite easy to comprehend fully. The contrast between physical infinity and nonphysical infinity is so sharp that it should be the basic assumption of modern science that we dwell in a universe where the appearances of matter can only be thought of as a presentation of nonphysical energy.

When it comes to the subject of earthly interactions with alien forms of life, it is possible to make an attempt at an enlightened rationality. Scientific fundamentalists have tended to dismiss the possibility of extraterrestrial visitors out of hand—none of the evidence that has thus far been presented has had any impact on them whatsoever. The rigid mindset of the fundamentalist type does not handle new ideas very well, especially ideas that do not possess the blessing of authority figures in their particular part of society. The subject of UFOs can be examined with an open mind, and an open mind is a rational mind.

Most of the sightings of UFOs and alien visitors have been made by witnesses who, in any other circumstance, would be considered sober, intelligent human beings. This is a critically important factor. In order to dismiss all UFO reports, it is necessary to dismiss all UFO reporters, and this is simply not reasonable. The fact that a wealth of detailed anecdotal reports by ordinarily reliable witnesses exists to be examined indicates that the phenomenon of UFO visitations is real, simply on this basis alone. A scientific fundamentalist who is not satisfied by one million sober witnesses would also not be satisfied by two million. It is absolutely impossible to pry open the mind of a deluded fundamentalist.

Of course, one aspect of the UFO phenomenon that gives scientific fundamentalists understandable pause is the fact that the aliens appear to have the ability to "pop" in or out of our reality at their whim. This sort of thing does tend to cause mainstream scientists to throw a fit because it would force them to expand their conceptualization of what is "real" and what is "not real." Also, the fact that these aliens apparently have capabilities to manipulate reality that far exceed our comprehension is the cause for intense insecurities in our best scientific minds. They do not want to face the fact that they possess an inferior intelligence to these alien visitors. The psychological aspects of the UFO phenomenon outweigh all others for the scientific fundamentalist.

Clearly, aliens have the ability to manipulate energy in a highly sophisticated manner. It can even be presumed that they have mastered the intricacies of inter-reality travel—from their own microreality to ours. There is a good chance that they come not from a different planet in our own universe but from a different universe altogether. A truly enlightened understanding of the nature of energy may allow for the breaking down of barriers between realities. It is this understanding that allows them to "pop" in and out as they wish. It is a neat, and scary, trick.

It is clear that these visitors have a fair amount of curiosity regarding us and have come here from time to time to investigate us. From the great number of reports that are available regarding their activities here among us, it is evident that they can be rather cavalier in how they deal with us—not always careful to avoid causing us pain and discomfort. This type of behavior does not necessarily mean that they are hostile; they are simply from another world. If they do indeed come from another reality, it seems quite silly for us to worry that they may have plans to invade and conquer us. Why would one reality be interested in the conquering of a completely different reality? That would not make any sense.

All we can do, until they choose to announce their presence publicly, is to continue to gather clues about their abilities and their intentions. Eventually, enough information will become available to reach informed and rational conclusions.

Early Creations

Perhaps a small amount of useful speculation can be made regarding the early developmental stages of the creative force that I term macrocreation. It can be presumed that the earliest efforts of this force involved mere rudimentary doodling—the simplest designs that we can imagine. It might be comparable to a child using crayons for the first time. A child can derive great satisfaction from learning just what is possible for him or her to do.

Even the most basic designs carry with them symbolic implications. A sharply jagged line, for instance, carries with it the meaning of anger or hatred. A curvaceous and sinuous line represents sensuality. A circle can indicate the concept of completion. Triangles have an inherently mystical quality. These basic forms can serve as the basis for a symbolic language that has fundamental importance.

This foundation of symbolic language was instrumental in the evolution of the macrocreation force. The evolution of symbolic language took place alongside the evolution of creative design. With each and every new flourish of design that was achieved, the macroconscious acquired an increasingly adept ability to express itself. Eventually, the realization of this universe was within its grasp.

Before the concrete extreme of this earth microreality was achieved, the macroconscious apparently indulged itself in a vast display of abstract expression. This universe that we perceive is substantially abstract in its nature—gigantic galactic swirls, white-hot starburst suns, black holes of an impenetrable essence, string particles, "dark" matter, and curved space. It would appear that this abstract expression may serve a primal creative drive, and our universe is the way it is because of this.

It can also be understood that this microreality stands as an island of concreteness and that human beings are the ultimate representations of concrete thought and expression. Most humans simplistically assume that our "physical" form of concreteness is the norm, but the fact that the bulk of universe has an abstract nature indicates otherwise.

Scientists do at least admit that the totality of nature's design does lie outside their full comprehension. But the fundamental principle that should be recognized is that nature **is** a design and, thus, a very strong indication of an overall conscious design. It is understandable why most scientists have ridiculed theological explanations that attempt to describe the "creator" of this design, due to the quite fanciful nature of these explanations. The easily apparent absurdities of theological belief systems have blinded most scientists to the fully rational evidence that conscious design does underlie our existence.

Although we human beings can take great satisfaction in the rich, meaningful concrete experiences that this earth reality has to offer, it is helpful for us to make some effort to see beyond the confines of this artificial construct in order that we may catch a glimpse of the abstract nature of existence.

To a certain extent, earth reality has been a "prison" of sorts for us, and we, the prisoners, have grown quite comfortable with the limitations that have been imposed us. An occasional "sneak peak" of the abstract will serve to remind us of the reality that we came from and the reality that we shall someday return to.

Creative Evolution of Earth

Charles Darwin observed that a process of evolution was at work in the natural world around him and proposed an entirely mechanistic philosophic system to explain it. Logical flaws in this somewhat simplistic approach have been noted ever since his writings on the subject were published.

I believe that all the available evidence suggests that the underlying force that propels existence is a creative force and not a mechanistic one. What thrives and survives in this earth reality is not so much the "fittest" as that which is the most creatively satisfying. Those aspects of evolution that can be described as mechanistic are there to serve the creative impetus.

One demonstration of the creative nature of evolution is to look at the Age of the Dinosaurs. It is worth noting that children are the ones who tend to be especially drawn to the world of dinosaurs because the instigation of that epoch springs from particular child-like sensibilities of the macroconscious. The child in us responds to the child in the macroconscious.

The "Godzilla" era was a creative phase of design and development—it was something that the macroconscious very much wanted to experience. This was an era that was truly "fun" for the macroconscious—fantastic creatures running rampant in creative abandon. The macroconscious apparently had playful creative urges that needed to be manifested. The Age of Dinosaurs was a unique time that offered unique creative pursuits.

Eventually, we can see that these particular creative urges were satisfied, and it was time to move on to a new developmental phase. The Age of the Dinosaurs came to an end, possibly rather abruptly. Some scientists have speculated that a large meteor struck the planet, which then kicked up such a thick cloud of dust that it obscured the sun and the global climate subsequently cooled. Whatever did occur at that time, it was a mechanistic device that served the creative need to bring the dinosaur era to an end.

We should realize that, because the vast majority of life forms that live in this world are nonhuman, it should be evident that the macroconscious finds great creative satisfaction in these nonhuman forms of life. There is no logical reason for human beings to assume that we are the most important aspects of creation. It was once an incorrect human belief that our earth was at the center of the universe. It is our human nature to want to put ourselves at center stage, but there is nothing rational that underlies this belief. Humans represent a very small part of creation—it would be to our benefit to recognize this.

One aspect of evolution that can be considered in a new light is the Darwinian concept that humans evolved from the lower primates. The macrocreation philosophy position is that all forms of life evolved from ideas, and not as a result of a mechanistic process. Humans came into existence for the same reason that aardvarks did, because it was creatively desirable. There is no question that there is a strong genetic link between humans and apes, but this genetic link is indicative of the underlying creative link that exists between the two species.

All forms of life in this microreality will maintain their presence here so long as it serves a vital creative purpose. When that purpose has been fulfilled, as was the case with the dinosaurs, creative evolution will enable a mechanistic process to accomplish what is needed. If the Age of the Dinosaurs had not come to an end, then their dominance over the environment would have stifled the growth of other forms of life, including humans.

While it is correct for us to be concerned when an animal species reaches the point of extinction, we should also keep in mind that there is some underlying creative rationale for any particular extinction. All species should be valued highly for what they bring to this world, but we should also remember all life is part of a grand design.

One related subject area to the study of creative evolution is the realm of cryptozoology. Most scientific rationalists tend to automatically dismiss the possible reality of these crypto-creatures because their presence in our world is so difficult to pin down. Perhaps the reason for this is due to the nature of these creatures—perhaps their presence in our reality is not a constant one.

In considering the Loch Ness Monster, for instance, sightings of this creature go back for generations. If even one of these numerous sightings is valid, then this means that the monster is an actual creature. The fact that the sightings of it are so very infrequent may be due to the "in and out" nature of the creature's existence. Perhaps there is some quantum aspect to their biological structure that allows them to be in two places at once.

This kind of speculation may be easy to dispute, but scientific fundamentalists have proven themselves unwilling and unable to speculate at all. A rigid mindset is not a rational mindset. A little imagination needs to be combined with a lot of logic in order to get a full grasp of the true nature of our reality.

The Environment

In recent centuries, human beings have had the creative opportunity to run rampant over the natural environment. People have pursued whatever developmental urges came to mind, generally with little or no regard for the negative impact on the environment that might result. We have denuded forests, damned rivers, reshaped the seashores, and have allowed our urban areas to become sprawling behemoths. Due to this mostly unrestrained activity, we humans have created an environment that appears to be in very real peril.

The macrocreation philosophy holds an alternative view of the environmental situation. One thing that should be noted is that it has only been in fairly recent history that human beings have been in such a domineering position as regards the natural world. Prior to the Industrial Revolution, humans were completely dominated by the environment of this earth reality, and this had been the case for our entire human history. For countless thousands of years, humans did not have the means to impact the natural world on a large scale. Now we do, and this represents a sharp creative shift from earlier times.

Although the clear appearance of things is that it is within human capability to completely destroy the environment around us, I do not believe that we actually possess that power. The entire earth reality is a designed creation of the macroconscious. For the recent industrial era, humans have been permitted to pursue the most aggressive endeavors that we can imagine for the sake of our own creatively expressive needs. Despite our pernicious delusions of grandeur, human beings are not in charge of the universe.

Of course, there is no question that the criminally careless, selfish, and greedy acts of human beings have brought death, brutality, suffering, and inexcusable ugliness to many aspects of the environment. Since the 1970s especially, people have come to recognize that our developmental urges can have dire consequences for the environment. Even though the entirety of the natural world is not truly threatened by the unforgivable activities of human beings, the sobering appearance that our actions create should give us cause enough for concern.

For creative reasons, it can be presumed that this era of human destruction was in someway necessary as a developmental stage for earth reality itself. The twentieth century hopefully represents the climax of this stage of human stupidity. The twenty-first century should likely see a gradual lessening of the infantile human desire to dominate the environment. This new century should bring humans to a point where a very solid appreciation of the natural world is now deeply ingrained.

Human intellectual understanding is always in a state of evolution. Thus, the "sins" of the past need not be carried over into our future. Environmentalists have been quite correct in becoming alarmed at the sad state of affairs that human actions have brought us to, but it should always be kept in mind that the destiny of the natural world is not in grubby human hands. The thing to do is to act as if this earth environment was our domain even though it is not—For the sake of human evolution, we must acquire a full measure of respect for the world around us.

We human beings are simply playing the role of sovereigns in this dramatic earth reality—the true power lies with the macroconscious, which has a full appreciation for all forms of life that are found here. We are not God's "children" any more than a raccoon is. Despite our unbridled and absurd dementia of vainglory, we human beings have not been given dominion over the earth, or anything else.

Individualization of Consciousness

The reason only a human being can become a Marilyn Monroe is due to the individualization of energy. Humans possess the highest degree possible of individual energy characteristics and therefore have the greatest potential to manifest uniqueness. One aardvark will generally be quite similar to another aardvark because animals are beings which incorporate a fuller degree of generalized energy aspects. This is the reason why no aardvark will ever be a Marilyn Monroe.

The thing that makes a human being human is the quantity of individualized energy characteristics that we embody. Individualization of energy, as it relates to all forms of life, can be thought of in terms of a hierarchy, with humans at the apex. This fact in no way equates to any manner of human supremacy over other forms of life, it simply means that we humans have the greatest capacity to demonstrate unique aspects of energy.

There is a price to be paid for this high degree of individualization, however. Of all forms of life, we have the weakest overt connection to the macroconscious. Our high individualization equals high isolation. The more individualized a life form is, the less it will be able to share with other forms of life and with consciousness itself. Because of this, human beings lead a very lonely existence, cut off from each other and from life.

In some ways, animals have it much better than do human beings. Animals can never be as fully separated from each other as humans can be separated from one another. As we can all readily observe, animals cannot be as completely distinguished from each other because they share so many common energy characteristics. This does not mean that each animal does not possess a degree of unique personality traits that makes each one special—every animal lover knows how each of their animal friends is truly distinct and like no other. However, it should be clear that the degree of an animal's uniqueness cannot match the degree of uniqueness that can be displayed by a human being. Once again, it needs to be emphasized that none of these factors of energy individualization makes an animal in any way a lesser being than a human.

Because animals have an overt connection with the macroconscious, it can be said that animals are much more enlightened than humans, who generally do not have a clue as to the fundamental aspects of this world around us. All animals are "geniuses" because they possess knowledge and understanding of this world that humans do not.

To their great benefit, animals are much more in touch with the creative nature of this world than are human beings. Instead of falling prey to the lifelessly mundane reasoning that science has to offer to explain this world around us, animals recognize the dramatic aspects of this reality for what they really are. For instance, daybreak at dawn and twilight at dusk are seen as fundamentally creative "art" by animals and not as the mindless workings of astronomy. The mechanistic physical "laws" that humans rely on so heavily mean very little to animals, who see the world for what it is—a "magical reality."

Of course, animals such as aardvarks and raccoons make up only a portion of the totality of life forms in this world. While human beings generally show a measure of respect for most animal life, when it comes to the insect kingdom, it is usually a different matter. For example, if an ant crawls across our carpet, we seldom feel much remorse when we kill it. Even if we kill a thousand insects who may have made a home for themselves in our back yard, we may not feel any guilt in destroying them. Instinctively, humans understand that to kill one or more insects is not equivalent to the destruction of an animal's consciousness.

When it comes to characteristics of individualized energy, insects tend to manifest very little. Because of this factor, the nature of insect consciousness is quite diffuse and generalized. Insect consciousness is a vast form of consciousness—a single ant is like a single cell in a mammal. To destroy a cell in a raccoon does not destroy the raccoon's consciousness because the consciousness is obviously much bigger than any biological component of the animal. Therefore, when a person steps on an ant, he does so "knowing" that the ant organism still lives on.

While humans exist in "splendid isolation" cut off from each other and from the macroconscious, insects are intimately connected with each other, as any biologist can attest. "Lowly" insects are actually grand creations, having a massive consciousness that dwarfs our human ability to comprehend. What humans label as "instinct" can be better understood as **comprehension**. Insects and all nonhuman forms of life comprehend this world in a way that is beyond our reach.

The more individualized energy characteristics a being possesses, the less comprehension that being possesses. This fact means that humans exist on the bottom rung of this particular hierarchy. We live almost utterly in the dark. There is no excuse for our grandiosity.

Over the years, there have been many wondrous reports of pets who become separated from their owners who then are able to locate their owners in territory that is wholly unfamiliar to them. This type of animal behavior is usually quite inexplicable to the traditional scientific mind. They do not want to believe that it is possible for an animal, or other nonhuman form of life, to have greater comprehension than human beings as to the workings of this world. Humans have for so long put themselves at the very pinnacle of biological existence that proper respect has never been given to any other life form.

Any time that an animal acts upon "instinct," it is actually accessing the macroself in a useful way. The vast knowledge the macroself has to offer is the explanation for all the extraordinary acts that animals and insects are so capable of performing. Nonhuman forms of life comprehend this world, while human forms of life, for the most part, do not. We humans generally pursue life in utter confusion.

Another form of life that can be looked at in a new way is the microscopic. Viral and bacterial forms of life are possibly the most truly grand in all existence. Their consciousness may be as vast as the ocean. It should come as no surprise that, upon more than one occasion, these tiny forms of life have completely outsmarted futile human attempts to control them. It is quite likely that microscopic forms of life have the ability to completely outmaneuver human beings and that whatever victories we think we may have over them, the victories we have are simply the victories we are permitted to have.

When it comes to the true reality of consciousness, we humans are the puny ones. We laud ourselves for our presumed primacy in the natural world, yet it is reasonable to assume that human beings could be easily wiped out if the microscopic forms of life made it their business to do so. Humans arrogantly consider "planet Earth" to be our world, where other forms of life exist at our discretion. However, given the magnificent scale of consciousness that the microscopic life forms manifest, it is much more reasonable to state that we live in their world and at their discretion.

All forms of life are intimately connected. We are connected in terms of consciousness and of energy, as well as all the more overt biological and ecological ways. Simply put, existence itself is possible because beings relate to beings and energy relates to energy. All the infinite designs of energy that have been wrought have served to create forms of life that have complex interrelationships with each other. The drive of the macrocreation force is to create energy relationships, possibly because these relationships have a fundamentally emotional characteristic. This quality of emotional relatedness may lie at the very heart of existence.

For humans to imagine that "God" is one of us is the grandest of grand illusions. It must be understood that the macroconscious takes great creative satisfaction in all aspects of created consciousness. While humans do play a unique role in the scheme of things, there is no rational reason to aggrandize this role. In fact, because human beings have less comprehension of this world than any other form of life, a full measure of true humility is called for. If we humans can learn to truly respect other life forms, we just might learn some important things from them.

The Energy Landscape

Applying the ideas of the macrocreation philosophy to the world around us can lead to the conclusion that every single spot on the landscape has a unique energy signature. This means that the landscape that we dwell in is an energy landscape, and each one of us has a meaningful energetic relationship with the little piece of land that we call our own.

An individual who spends his entire lifetime in the same place—Kalamazoo, Michigan, for instance—will have a personal energy matrix that has been strongly influenced by this situation. Kalamazoo, Michigan has an energy matrix that is unique to it, and this energy field will energetically influence all who reside there. Thus, every person is affected by where they live and how long they live there.

As with all relationships of energy, the nature of these relationships can be described as sympathetic, neutral, or antagonistic. A person may find that they flourish in one locale but wither away in another, for no reason other than the energy interchange between person and place. Most of the time, it appears, the energy relationship between a person and a place is fairly neutral, which makes it generally easier for us to move around and relocate from time to time. Also, it should be said that the energy relationships between a person and a place may be rather complex—it may have both "positive" and "negative" aspects. The nature of energy relationships should never be over-simplified.

The obvious distinctions found on the surface of the land relate to the underlying differences of characteristics of energy. A desert landscape will have an energy matrix that relates to all the visible attributes of a desert while swampy places will have a comparable swampy energy aspect. Those qualities that make New Orleans, Louisiana so distinct will be the innate qualities of energy that it possesses. These qualities of energy will serve to attract individuals who have sympathetic energy matrices and repel those people who have antagonistic ones.

Most of us have had the sensation of arriving at a locale for the first time and having instantaneous reactions to that place. Sometimes, we will be strongly attracted to a city or a rural area for no discernible reason—it simply "feels" good. On other occasions, a new place will almost immediately repel us, feeling inexplicably "bad" or even "evil." Energy interaction is one possible explanation for these feelings that we have not been able to understand.

There are times in our lives when we will be drawn to a locale that will have certain stimulative aspects because stimulation is exactly what we need at that point in our lives. At other times, a place of very low stimulation may be what is required. Because of the intricacies of the energy landscape, some new locations will immediately feel "homey" and comfortable, while others will seem "cold" and uninviting. The more we travel, the more we can tell that this is so.

The longer we remain in a particular locale, the more ingrained will be the relationship between place and person. We are who we are because of where we are. Those people who move away from their hometown after high school or college graduation and then return twenty years later often notice that a gulf has developed between them and their hometown friends. There are obvious reasons for this phenomenon besides energy, of course, but I believe that the energy connections between person and place explain some of what is going on.

An individual who resides his entire life in small-town New England, for instance, will take on the energy aspects of New England to a quite significant extent. This person, with every word and gesture, will embody the characteristics that we associate with that region. Alternatively, the person who rambles ceaselessly around the country will often take on an enigmatic "air" and become someone whose nature is rather difficult to pin down. The rambler, for reasons of energy complexity, will be someone who is "everyone and no one."

It can be theorized that large cities spring up where they do—in addition to the obvious factors of their locale—due to the energy potential that is inherent in that area of the energy landscape. A particular locale can be, in terms of energy, a city waiting to happen. Looking back in history to the development of the United States in previous history, it is easy to see that "destiny" favored some spots and disfavored others. Chicago, for instance, in its early days was a town no bigger than dozens of others in the region, yet it was the place that exploded into the behemoth we know today. This type of explosive development of one spot over another may be considered an utterly random event by many, but it is possible to look at urban growth in the context of energy and conscious design.

The borders that we see on our maps may appear to be completely arbitrary in nature, yet I believe that these "superficial" borders represent inherent borders of energy. Any nation, state, or other legal entity can have an energy signature that is unique to it. Although many characteristics of the states of Illinois and Indiana are highly similar, those people who know those states well can tell you that one state is truly different from the other, if only as to how they "feel" to the resident. Outsiders may not have a clue as to these differences, but they are there.

Beings composed of energy have inescapable relationships with a landscape that is also composed of energy. Existence itself is defined by the relationship of one aspect of energy to another—this is really the fundamental essence of life. We can alter our personal matrix of energy by changing our location—but we also strongly affect our energy matrix by simply staying put. Places define us, and we define those places. Energy defines all.

Human Cultural Evolution

Because energy can be considered the essence of all things, it is reasonable to conclude that there is a direct and comprehensible relationship between energy and time itself. This would mean, for instance, that every minute of every year would have its own unique energy signature. Applied on a grander scale, this would indicate that each year also is unique in its energy patterns—the year 2000 is simply a profoundly different energy matrix from the year 1900, for example. Using this line of thinking, human culture can be looked at from a truly fresh perspective.

We can readily see that each new human generation that comes along appears to be, in some ways, different from their parents. Each new generation represents an energy evolution of the human life form. This is certainly most in evidence in fast-paced modern societies where each new generation ushers in dramatic new changes. In slower-paced cultures, the evolutionary changes from one generation to the next may be much more difficult to observe, but very possibly they are indeed there.

People born in the year 2000, for instance, will be oriented, as a rule, to the energy signature of the year 2000. Each new generation embodies a unique set of energy characteristics that will never be repeated. It can also be assumed that each generation will manifest certain energy traits that will serve the creative designs of the macro level of consciousness. For an example, the generation that came of age in the 1960s—in many nations of the world—demonstrated strong revolutionary characteristics that were not to be found in any recent previous generations. These "radicals" of the 1960s had what was needed, in terms of their personality traits, to take this world in vibrant new directions. One could make the contention that all aspects of human cultural evolution are completely random, but I believe that a more logical argument is that major events in our society, such as the revolutionary 1960s, occur as a result of macroconscious design.

It is creative impetus that motivates human society, and not incoherent randomness. The "Baby Boom" generation stirred brand-new currents in nearly every aspect of life—the art world, the music world, literature, cinema, television, advertising, sexual issues, drug experimentation, politics, and philosophy. These earth-shattering changes came about as a result of design goals, accomplished through the efforts of a new and unprecedented generation. The Baby Boomers possessed energy matrices not unleashed on human society before, and this is how the world became a place where new things could happen and new directions could be taken.

At the beginning of the twenty-first century, the world appears to be dominated by computers and by people who love to use them. Perhaps the present generations are particularly attuned to the digital world in a way that previous human beings were not, and this fact has allowed computer programs to dominate human culture in the overwhelming way that they have. It could be said that computer applications dominate us within and without.

It may be a facet of energy that the energy configurations present in a child at the time of his birth will remain with him for the rest of his life, and this will affect his ability to accommodate himself to a future world by the time he grows old. This is to say that a child born in 1960, for instance, will always be, in a significant way, configured to the energy patterns of the year 1960—it would be inescapable. As the child ages, an increasingly large "gap" will develop—the time configurations of the world will continue to evolve while the individual remains "time stamped" at 1960. By the time the child has reached senior-citizen status, a major time displacement will have occurred.

Every nostalgic adult understands very well that the era of his youth has a certain potency and meaningfulness that has eroded as time has passed. Because of the relationship of energy that exists between people and time, an individual will be much more in sync with the world that existed at the time of his youth and young adult years and much more out of sync with the world he finds himself in during the latter years of his life. I believe that the common sensation that we notice of time seeming to go faster as we get older can be related to the changing energy relationship that we have with the world around us—a world that is out of sync is a world that does not capture our attention nearly so much. To a real extent, we are less "in" this world the older we get. The world can never again be as potent as it was when we were young and "in tune."

None of this means that senior citizens are doomed to an empty life of endless nostalgia. We all know that aging presents us with many challenges, and one of these challenges is doing what we can to give as much attention as we can to the world that currently exists around us. The world that existed in our past will generally seem like a much better place than what we find around us today, and so extra effort on our part is required to keep things relevant. The easy thing is to be sentimentally nostalgic—the hard thing is to find meaningful things in a modern world that may feel harsh and unpleasant in many ways.

In some instances, it appears that certain individuals have an energy orientation that links them to a time period other than the one in which they were born. For example, a child born in the year 1960, for example, might not necessarily be completely oriented to the year 1960—it might be 1860, 1442, 150, or any other era. This orientation to a particular period in the past is helpful because it serves to maintain an ongoing link with previous time frames. If every single person born in a certain year was oriented solely to that year, then it would be much more difficult for human cultural history to be brought forward to the current time.

Perhaps all human beings have some energy link to other time frames, whether it be in the past or in the future. But, with certain individuals, this link can certainly be considered quite strong. With most of us, the connection to another period of history is something that we can take note of, and take pleasure in, but it does not overwhelm us. Some of us experience a poignant nostalgia for a period of history that has a significant influence in our lives, and this idea of energy connections is one way to explain it.

It also happens that certain individuals will be strongly oriented to a future time frame, which will put them "ahead of their time" in various ways. This energy link to the future is obviously helpful in moving society forward to new achievements and endeavors. Inventors and other visionaries are undeniably "forward-thinking" individuals—their work helps move society ahead to a future that they can already see. This is also true of important political leaders who truly do fulfill the responsibility of leading human society to a more evolved plateau. The old saying that it takes "all types of people to make up a world" can be understood to mean that it takes all kinds of energy orientations to achieve a society that is properly balanced.

Most scientists look at time, and everything else, in simplistic mechanistic terms. They assume that time, under normal earthly conditions, unfolds at an unceasingly steady pace. Looking back at human history, however, it seems quite clear that some eras have been rather quiescent while others have been incredibly active. It can be questioned whether time proceeded at exactly the same pace in the twentieth century, as it did in the twelfth. The twentieth century was a creative maelstrom, with each decade ushering in an amazing new era of human culture. Human society does not proceed at a slow and steady pace—there are occasional eruptions breaking up periods of dormancy. Traditional scientists have had no ability to explain the creative nature of human society.

The fads and fashions that we observe in our popular culture can be attributed to the energy that the macroconscious invests in them. Each new era that we witness is like a new "show" created for our benefit. The macroconscious is the auteur of this world stage. New music, new fads, new art movements, new developments in stage and screen—all these are designed and orchestrated by the macroconscious using the power of the macrocreation force. The Beatles, Pablo Picasso, Tennessee Williams—every human artist is at the service of creative maneuvers implemented at the macro level.

The musical era that was launched by the Beatles provides a good example of creative macro design. Millions of young people around the world responded to the literal jolt of energy that the Beatles provided with their music in the 1960s. The world had never seen anything quite like it—pandemonium breaking out due to the well-crafted songs of sweet love that the Beatles had to offer. This music energized a generation, setting young people off in new directions that impacted society on a broad scale. None of this was sheer randomness—it was conscious design.

In the 1960s, society was effervescent with infused creative power—pop art went psychedelic, hairstyles went wild, fashions went mod; the world changed and evolved in a lightning fast manner. The fervent energy of the 1960s radicalized many political beliefs, and people took to the streets all over the world. The decade of the 1960s displayed raw creative power that stunned humanity. There might never be another time like it.

The macrocreation force can be invested in any aspect of existence, including individual human beings. The Beatles were able to impact popular culture the way they did because the requisite creative power was invested in them. On a much smaller scale, many actors, singers, bands, composers, comedians, dramatists, and others have had successful careers that impacted society to a meaningful extent because energy was invested in them that made those careers possible. However, this creative energy can most certainly be withdrawn from any individual at any time, thus resulting in some manner of career death for that celebrated person.

We have all seen comedians or comedic actors who simply "stop being funny" after a few years—whatever made them hilarious at an earlier time has gone away. In other cases, an actor or painter or writer may accomplish something truly brilliant at one point in their career and then be unable to duplicate that success. These people did not cease being "brilliant" as such; they simply were required to fade away in order to make room for the newcomers who were waiting in the wings. For purposes of creative maneuverings, a star may be on the top of the heap one moment and washed-up the next. This is how creative energy in our dramatic reality is made to work. It may seem cruel at times, but it is a creative necessity.

In popular music and in other creative endeavors, the twentieth century certainly appears to be the ultimate expression of creative infusion from the macro level. Simultaneously or in quick succession, there was the Jazz Age, the Big Band era, blues, country and western, rock and roll, folk, disco, new wave, punk, grunge, rap, and numerous variations of these forms. The creative growth seen in literature, cinema, television, radio, and other artistic arenas is staggering to contemplate. The scope of creative achievement seen in the twentieth century is so mammoth that it may just be that this recent century is the focal point of human civilization—everything in human culture led up to the twentieth century and, now, everything will lead away from it.

Despite what many people believe, "genius" is not a gift that is bestowed as a favor by a mythical God. People who can be labeled as geniuses are those people who are serving as useful instruments of energy, accomplishing various creative goals as an aid to society. The creative brilliance of Pablo Picasso, for instance, re-directed the development of modern art. The evolution of art would not have occurred unless there was a human instrument in a position to make it happen. Without this occasional manifestation of "genius," the world would be a stagnant place, simply repeating the efforts of the past.

It was the combined genius of many individuals that has brought us to the computer age, an era that represents yet another major evolutionary step for human culture. The advent of "virtual reality" has become an accepted idea that will have profound philosophical implications. The nature of a programmed reality will be increasingly compared to the nature of our own reality—the similarities are too potent to ignore. A virtual reality is a reality that came into existence as a result of ideas—conscious design—and this fact will be making a deeper impression in many philosophical systems in the years to come.

It has been the traditional belief of most New Age groups to imagine that humanity has been engaged in a long, hard struggle for universal enlightenment—this is simply too pleasing a concept to easily let go of. If humanity is, in fact, on some sort of path to greater enlightenment, it is because "enlightenment"—however one defines it—is a more satisfying path in creative terms than continued darkness. A belief system, ideally, should never be based on mere wishful thinking, but on rational and coherent applications of logic. A New Age will occur if it needs to occur, and not because it is a nice idea.

Having said that, it does appear as if humanity is indeed engaged in a process of breaking away from its unbridled barbarism of the past. There is no question that frequent warfare still breaks out, but it seems as if we barbaric humans at least take the time to contemplate alternatives to war much more readily than in the past. War is not a universal pattern anymore; it is a calculated, isolated event that is generally agonized over by both the public and society's leaders. War is not a free-for-all anymore; it is a sober undertaking.

To point out the obvious, it can be noted that human society has achieved the status of a "global village" that has made us all interconnected economically and culturally. Nearly all the nations of the world are working together to achieve common economic goals and mutual progress. Because the macrocreation force demands incessant cultural evolution, it is guaranteed that the world we find in the twenty-first century will be significantly different from anything that has come before. It is not unreasonable to hope that current trends in global interdependence will continue to unfold in a way that will allow this world to be a brighter place than before.

While many positive statements can be made regarding the achievements of the twentieth century, that particular time frame may have also manifested the pinnacle of certain undesirable cultural characteristics as well. Perhaps the twentieth century represented the zenith of human greed, ethnocentric schisms, tribalism, militarism, religious intolerance, and xenophobic nationalism. If the previous century does display the acme of these human faults, then whatever comes next may certainly give the appearance of being a wondrous New Age of human development. There may be little energy invested in warfare by the macroconscious in the future because the full creative aspects of war have already been exploited. This cannot be considered a prediction; it is just a possibility.

What humans have been capable of doing to ourselves in the past we may be incapable of doing to ourselves in the future. This is not to say that there will not be some "dark" periods ahead—there most certainly will be—but a trend toward greater human tolerance seems to be very much in evidence. Humans grow and develop, both as individuals and as societies, because it is creatively satisfying to do so. Human beings like nothing better than to compare ourselves favorably to our ancestors, to applaud ourselves for just how far we have come. Because we all have a deep need to feel as if we are progressing, we will continue to progress. We witness our own evolution, and we are encouraged.

The Macroquotient

The macroquotient (MQ) is a means of understanding, in a shorthand fashion, an individual's personal level of awareness. It is a term that is meant to be used in a similar way to IQ (intelligence quotient), but the MQ has no direct relationship to IQ. The MQ is a "label" that can be attached to a person in order to provide a quick method to identify their basic nature.

An individual who can be readily described as intuitive, open-minded, and possessed with a fairly high degree of objectivity regarding his role in life is an example of one who can claim a relatively high MQ. Alternatively, someone who tends to be rigid in most ways, closed-minded, and quite subjective in his viewpoints can be considered a low MQ type. The macroquotient can be seen as one's conscious connection to the macroself level—the acceptance of knowledge and perspectives that come from a place that lies beyond our ordinary, everyday level of awareness.

Low MQ people generally will fully inhabit their role in this world—they live it and breathe it. Those who are born into a solidly defined ethnic, religious, or economic environment will most likely be fully defined by this one environment, and they will have little insight into how other people view the world. Fundamentalists of all stripes quite clearly fall into this low MQ category.

Because of their utter lack of objectivity, low MQ types may find this world to be a bit more vivid and dramatic than those who possess a higher MQ. For the low MQ individual, this world is composed of black-and-white aspects—all things can be instantly characterized as right or wrong, good or evil. This high degree of subjectivity can bring about a concurrent exhilaration that can act like a drug—a sharply defined world is a more exciting world than a reality made up of tones of gray. Fundamentalism is an intoxication of sorts, and it can be quite difficult to "kick" the habit.

Racists, sexists, xenophobes, homophobes, and others who live within a constricted mindset can be described as low MQ individuals. While some of these people may possess a very high level of intelligence, their profound subjectivity will usually smother their intelligence, and they are left with very little insight. The low MQ type lives deep within his definitions, barely able to see out of the hole he has buried himself in.

This quality of a low macroquotient can also manifest itself in entirely personal matters. For instance, a husband may discover that his wife has been unfaithful to him and his reaction may be to explode in a violent, jealous rage. This state of temporary insanity can occur because the husband has been unable to acquire any objectivity on the situation. With a higher MQ that allowed for a greater objectivity, the husband might have been able to see how his own actions had contributed to his wife's unhappiness. The low MQ type has little insight into his own motivations or the motivations of others. The result of this wholesome subjectivity will often lead to trouble and trauma.

In contrast, the high MQ individual is highly tolerant, reasonably objective about themselves, and intuitive as to the needs and desires of others. While all of these qualities are admirable to have, it can also be true that high MQ people may tend to be somewhat indecisive, wishy-washy, and prey to feel-good notions regarding this world. A high MQ person may perceive all paths in life so clearly that no single path stands out. Also, the high MQ individual may be so empathetic that he is unwilling to do anything that might cause the slightest pain to another person, even if that pain might be the best action to take in a particular situation. Too much objectivity can be a paralyzing agent.

Someone with high MQ characteristics may demonstrate telepathic abilities, clairvoyance, and other "psychic" strengths. This significant degree of sensitivity that the high MQ type embodies does not, obviously, go hand in hand with common sense. The high MQ person may be attracted to belief systems that promote "pie-in-the-sky" philosophies that serve mainly to make the adherent feel good or even "blissed out." Most certainly, the extremes of a high MQ and a low MQ offer great challenges that will need to be faced as best one can. Life is not easy for anyone.

It should be stated again that there is no link between the intelligence quotient and the macroquotient. In addition, most human beings manifest a complex mix of abilities and affinities that may defy a straightforward MQ characterization. In order for this dramatic earth reality to unfold as it must, there is a definite need for the low MQ types to play their roles as fervently as they do. For the foreseeable future, a balance of macroquotient aspects will be essential to the functioning of this world.

Human Energy Relationships

It is inescapable that an energy relationship of some sort will occur between any two people who come together or between any groups of people who have proximity to each other. What has been labeled "chemistry" in the past, as in romantic relationships, can be better understood as the action of one's energy patterns upon another. While many energy relationships can be characterized in simple terms—neutral, sympathetic, or antagonistic—most relationships between beings of energy will be sufficiently complex to involve a combination of these various aspects.

When two people encounter each other for the first time, a mutual energy assessment will take place beneath conscious awareness. Because of this, each person is able to "read" the energy signature of another and reach some instantaneous conclusions. It is not at all unusual for two strangers to become instant friends if they share a highly compatible energetic nature. Conversely, two strangers can also very easily "rub each other the wrong way" if there is strong energy antagonism involved. For most people, most of the time, a fairly bland neutrality will characterize most social interaction.

One major challenge that faces humanity is the inherent energy antagonism that can exist between groups of people, often racial groups. This energy antagonism that exists between large groupings of people is a design feature of this dramatic reality. Racial and ethnic hatred, for example, have very little to do with external features (such as color of skin) but instead relate to internal energy configurations. Physical differences between the races cannot really be used to explain the high level of conflict that has often existed in our society. Something deeper and more creative is going on.

The conflicts that we deal with between groups of people significantly add to the challenges we face in this earth reality, and this is the way it is meant to be. The kind of problems that we must deal with here are problems that relate specifically to this arena. Racial and ethnic hatred simply are not factors in the greater reality—macroreality. We face these situations here because we do not face them there. Our world is a world of earthly challenges.

The variety of human groupings that we experience here in this reality serve to provide a stimulating creative mix. If all groups of people were in energetic harmony, this world would exist as pure monotone, and that would be creatively unsupportable. Each racial, ethnic, or other grouping can be said to manifest a unique set of characteristics that is essential on a creative basis. This energy differentiation between groups should never be thought of as energy inequality. The idea of inequality is a human concept—it has no value when considering the nature of energy.

It is a sad fact that this energy antagonism that can exist between groups has been the instigating factor for countless horrific episodes of racial, ethnic, nationalistic, and ideological hatred. While energy antagonism is indeed a design feature of this microreality, hatred itself is not a design feature. While it is always an option for one person to hate another due to energy antagonistic factors, it must be said that **not** hating another person is also always a valid choice. Energy antagonisms present us with a challenge—mindless hatred represents a failure to meet that challenge.

We can do nothing to escape our energy relationships. There will always be compatible people and incompatible people in our daily lives. Most religions and New Age philosophies blandly urge us to love all other people as we love ourselves. However, this is not realistic given the powerful aspects of our energy relationships. There is no reason why we cannot avoid extremes of behavior when dealing with one another—we need to recognize the challenges that we face for what they are and deal with them appropriately. It is best if we can keep our expectations reasonable.

The idea of a utopian integration of all human groups into "one big happy family" is not a likely occurrence. However, a natural process of integration that takes place entirely as a result of freedom of choice is certainly something to be hoped for. Those people who promote goals of wholesale social engineering will always face disappointment—people are energetically incapable of jumping through the social hoops that governments may choose to set up for them.

Relationships between people and groups of people are highly dependent on adequate interpersonal communication, and energy can be seen as a component of this ability to fully communicate.

Communication takes place on two levels—the surface level and the underlying macro level. We talk to each other using words, which can be thought of as concrete representations of macro-level symbolic language. Human languages are superficial devices for communicating with each other—the essence of our communication can actually be found in exchanges of energy.

Alone among life forms, human beings utilize a concrete language as the primary means of communication. All other forms of life communicate perfectly well with each other without the use of concrete language because they can connect with each other through signals of energy—they know what they need to know from the other by accessing the macro level of consciousness. Also, of course, animals are quite adept at "reading" each other's body movements, as well as learning new information through their sense of smell. Nonhuman forms find concrete expression through words quite irrelevant, for the most part. Only humans reduce the art of communication to the superficially concrete.

All beings are linked with each other because we are all aspects of the continuum of energy that comprises existence itself. Macrocommunication is the interchange that takes place within this continuum. With human beings, we are generally unaware of the macrocommunication that takes place between us because we are so highly focused on the concrete languages that we all utilize. Many people consider telepathy, for instance, to be mere poppycock, just a fantasy that some delusional New Age types indulge themselves in. But energy interchange is not a delusion.

Macrocommunication involves direct knowledge. Those forms of life who engage in macrocommunication do not engage in lies, subterfuge, flattery, or half-truths—these things are usually limited to the human form of expression. Forms of life who use macrocommunication exchange knowledge with each other in an utterly direct way—there is no joke telling, gossiping, or hyperbole. These aspects of interchange tend to be a largely human indulgence.

Due to the significant limitations of human language, communicating with each other can be fraught with difficulty. This difficulty can certainly be magnified when strong emotions are involved. We shout at each other in an effort to be heard, yet may understand next to nothing.

We often say something with our body language that can be in conflict with the words that are coming from our mouths. In addition, we all communicate telepathically whether we have any knowledge of this or not. Everyone accesses everyone else at the macro level and processes this information; we simply choose not to acknowledge this to our conscious minds. Human beings usually like to keep things simple, but there is nothing simple about interpersonal communication.

Another concept that can be dealt with in regards to human energy relationships is the notion that there is such a thing as "male" and "female" energy. This simplistic idea has appeal for many people because it seems to offer a nice explanation for how certain things operate in this world. The macrocreation concepts regarding the energy spectrum will be discussed later, but what can be stated now is that the full range of energy characteristics cannot be neatly split into two categories.

Any individual's personality aspects are the result of the complex matrix of energy configurations that constitute a living being. Sexual characteristics, including sexual orientation, are the result of the energy matrix that is constituted at or before conception—it is a matter of program design. Increasingly, society has come to realize that its old-fashioned notions concerning sexual deviations from the norm have little or nothing to do with a mishandled upbringing but are due to biological aspects. But beneath the biological is the energetic.

The macrocreation philosophy holds that all major sexual tendencies that an individual can manifest are the result of energy design. This would mean, as an extreme example, that a predatory pedophile was programmed to be exactly that. Any individual with this sort of programming faces an immense challenge in life, and while society has the duty to do all that it can to protect vulnerable members of society from any predator, society should keep in mind that the predator is not an evil monster, but rather someone who is intrinsically aberrant.

For anyone in society to react with mindless hysteria to the actions of sexual deviants is simply not helpful in dealing with those deviants. One possible avenue of treatment is the use of virtual reality systems that would allow sexual predators some small measure of satisfaction in their lives. There can be no rational reason why this harmless outlet should be denied to anyone who might benefit from it. An attempt should be made by society at large to suspend moral judgments against those people who find themselves afflicted with a particular sexual nature that has no wholesome outlet.

The moral judgments that hold sway in certain societies today were conceived in utter ignorance thousands of years ago. These judgments have been upheld as "sacred" by some religions because the idea of tradition is valued so highly by so many people. However, if a theological position is based on irrational notions conceived by an odd bunch of fundamentalist zealots in the very distant past, this position should be considered ripe for review. Few would argue that the religious zealots whom we see around us today should still be listened to two or three thousand years from now.

Ultimately, any choice made in the sexual arena must be considered a creative choice, which means it cannot be completely dismissed due to its possibly unseemly nature. It is necessary that our laws continue to do what they have been doing—protecting us from each other when this is what is needed. But it is desirable that these laws be based on the idea of an energy program design rather than on theological principles from long ago. Society should proceed on the assumption that all manner of sexual behavior will find some manifestation in our society, and it will need to be dealt with in a rational manner. Sexual satisfaction should be thought of as a basic human right, and society needs to be as imaginative as it can in order to deal with all kinds of sexual desire in a way that will not be harmful to anyone. Since the sexual challenges that an individual faces and that society as a whole faces will not be going away, there must be effort made to handle these challenges in a rational and humane manner.

In the question of romantic love and powerful sexual relationships that people can experience, the maneuverings of the macroself must be taken into account. Every romantic relationship that takes place, whether it is lovely and glorious in its nature or disastrously destructive, occurs at the instigation of the macroselves of the people involved. The macroself, from its perspective, is quite aware of the creative implications of any particular pairing, even if the human self is completely oblivious. While a human being may believe that all they want out of life is a sweet, simple little love affair, the macroself knows better.

It should be remembered that the reality that we inhabit here is a dramatic reality, and drama is going to be a big part of our lives, whether we like it or not. There will be passion, and there will be pain because this world needs to fulfill its creative function in the greater scheme of things.

The macroself can work to guide the human self into a romantic love situation as well as out of it. Emotional responses can be fairly readily manipulated at the macro level. A love affair can be seen as a creative affair, not at all dissimilar to an artist dabbing at a canvas or a composer writing a song. If a certain level of emotional pain will be in certain service of the overall creative design of the macroself, then it is necessary that the human self experience this pain. Love can hurt, and this hurt cannot always be avoided.

To sum up, human energy relationships are the result of a complex array of energy characteristics that are utilized in a creative program design. Whenever we encounter any sort of difficulties in our relationships, we should be aware of this creative design, and this will help us to acquire some degree of objectivity in most situations. We should also be aware of the designed energy antagonisms that can exist between individuals and between groups of people. This knowledge should help us to cope with the inevitable challenges that we will all be facing as we live out our lives in this highly dramatic reality.

Morality

As is the case with everything in the world around us, morality can be understood in terms of energy, and by doing this, remarkable new conclusions can be made regarding how morality truly operates in human society.

The moral systems that human beings employ today are entirely human-centered ones and cannot actually claim any degree of "divine" approval. Our moral codes generally date back thousands of years and have gained the measure of respect that they have in our society more for reasons of esteemed tradition rather than for more worthwhile reasons. These ossified codes can still, in some cases, serve society ably. In other instances, they do not serve society well at all. These moral systems that humans employ do not automatically correspond with the moral system found in macroreality—the greater reality.

It is not accurate to describe the macroconscious as being amoral, yet what a human being considers to be right or wrong may have no relevancy whatsoever as far as the macroconscious is concerned. Despite what some religions proclaim, no one is ever consigned to "hell" for his misdeeds, and no one is ever rewarded with heavenly rapture for his good deeds. This is not how macroreality functions. The traditional concepts of heaven and hell are fundamentally human ideas—they do not come from "on high."

Human moral systems can be considered essentially mythic in nature. People, throughout the ages, have dreamed up a wide variety of moral imperatives and then foisted these moral directives on the mythical figure of God. Numerous "servants" of God have felt quite comfortable in proclaiming to a gullible populace their idea of what God thinks and what God wants. Far too often, human beings have shown themselves willing to believe that certain charismatic people who make the bold claim of speaking on behalf of God actually do possess this ability. This human tendency to want to be led is quite understandable, but it is something that should be resisted.

While the moral systems that human beings have concocted have no "divine" validity, they do at least address very real issues and concerns. Evil itself is not a mythical concept, nor is benevolence. In fact, what we label "good" and "evil" can actually be understood as attributes of energy. Anyone who has lived for a significant amount of time has encountered both good and evil—even a rationalist atheist cannot deny that there are good people in this world as well as bad people. An atheist has no explanation to offer as to why these attributes of behavior should be manifested in our species. Atheists recognize traditional religious ideas for the mythical systems that they are, but beyond that, they are stumped.

Scientists have observed for a long time that energy can be understood in terms of a spectrum. The color spectrum that is found in radiant energy—light—is the best known example of an energy spectrum. In addition, there is also the chemical spectrum and the electromagnetic spectrum. The macrocreation philosophy contends that the entirety of energy can be seen as manifesting in various spectra. Since everything in existence can be considered to be aspects of energy in myriad forms, even such a subject as morality can be looked at in energetic terms.

It is possible that all the many traits found in human behavior can be seen as attributes or a moral energy spectrum. For ease of understanding, this moral energy spectrum can be compared with the color spectrum that is found in light. As an arbitrary example, a shade of deep purplish red could relate to what humans label as "evil" in our society. This particular shade of color would then be represented in our society alongside all the other colors that comprise the spectrum. This simple idea can be used to explain why evil exists in this world—it exists because it is intrinsic to the complete expression of the energy spectrum. There is thus no way to avoid it.

A human being's personal energy matrix would presumably contain a full array of energy "colorations," with the actual expression of these "colors" varying significantly from person to person. An individual might embody a high degree of this evil shade of purplish red, or there might be very little. Also, that evil shade of red would have a unique configuration of other "colors" of energy to interact with that would differ in each case. Every single person presumably manifests the complete range of this energy morality spectrum, but no two people would share the exact same energy configuration.

Someone in our society who can be readily labeled as "evil" is someone whose individual palette of energy attributes have led to a definite result. This would mean, in essence, that a person can be "born evil," which is a very unpleasant conclusion to draw. However, objective observation of the lives of the infamously evil very often show that their evil revealed itself very early in their lifetimes, before their environmental circumstances had a chance to impact them much.

For there to be goodness and benevolence in this world, it is necessary that there be evil and selfishness as well. In order that any aspect of the energy spectrum be manifested, all aspects must be manifested, and this means that some people will possess to a strong degree moral aspects that society will consider to be unacceptable. An "evil" person is that way so that everyone else can be who they need to be. This fact does not excuse evil deeds that are committed by individuals, but it is helpful if society can have some real understanding of the challenges that these people face.

In this world, we cannot pick and choose the aspects of energy that we wish to see expressed—they all need to be expressed. It is wrong to characterize human society as being engaged in a battle between good and evil, because no battle is being waged. In the color spectrum of radiant energy, blue does not do battle with red. We understand without any confusion that red and blue are aspects of the greater whole—every single color and shade have their place in the color spectrum. None of these colors can be eliminated. In this world, we cannot escape evil any more than we can escape good.

These factors do raise the question of proper punishment for evildoers. While there may be a rational reason for excusing their misdeeds, in order for society to function as it should, laws must be enforced and evildoers must be punished. It is probably best if we acted as if evil were not a functional necessity—we will still need to enforce those laws that serve to constrain evil from harming us all. It is desirable, however, to respond to criminals with a measured degree of compassion, because we do understand the great challenges that they have faced dealing with their basic natures. There is every reason to treat criminals humanely.

Every individual in this world has a unique set of energy characteristics that will evolve in its own way, and no person's life is plotted from beginning to end. It will often be the case that certain energy traits will come to dominate someone's personal evolution—thus, a person who has some criminal tendencies may eventually come to be dominated by those evil tendencies. Their lives were not predestined to be criminal, as such, but they possessed all the requisite energy traits that could make a life of crime become a distinct possibility. The path that a person takes through life is composed of moment-by-moment choices, and these choices, no matter how seemingly insignificant they are, actually are of great import. This particular topic will be dealt with at length in another chapter.

Religious belief systems are not a necessity when it comes to leading a morally upright life. As a rule, atheists and agnostics can be depended upon to be as morally rigorous as their religious brethren. People live moral lives because it feels like the right thing to do, regardless of mythical portrayals of heaven and hell. We generally notice the quite immediate satisfaction we derive from doing the right and responsible thing as opposed to the wrong and irresponsible thing. What fate awaits us after death is almost always a vague feeling. We live for the way we feel right now.

The energy spectrum requires that the totality of the spectrum be manifested in this microreality, which is why we see the wide array of human behavior characteristics that we do. We need to understand the creative necessity of evil while at the same time doing all that we can to guard against it. Those who are brutal need not be treated brutally. The "Golden Rule" is still a wise rule to follow, and it cannot really be improved upon, even in a macro context.

Quite obviously, animals do not need human-based moral systems in order to live lives filled with generous love and a strong sense of duty. Animals are sufficiently tuned in to guidance received from the macroconscious that an "artificial" system of morality would be utterly absurd. Animals simply know what is acceptable behavior and what is not. They do not need to spend one single minute wondering what God wants them to do.

For humans, however, since we are cut off from this direct macroconscious level of guidance, artificial moral systems are unavoidable. It should be understood, though, that the moral systems that we invent for ourselves do not necessarily have a direct correspondence with the moral system employed in macroreality. This, too, is a subject that will be addressed later.

Villains

There will always be villains among us. Because it is necessary that the "evil" aspects of the energy spectrum be manifested in this dramatic reality, there need to be people among us who will embody these energy aspects. Villains play a crucial function in this created world—they are quite essential in the unfolding of dramatic events in our own lives and on the world stage as well. A thrilling movie or novel requires a memorable villain to set the action and motivate the hero. We can justifiably detest villains for the damage that they do, but we must acknowledge the vital role that they play in earthly affairs.

This world is designed to be a realm of intensity—it is desirable that we become deeply involved in the events that transpire all around us. Tyranny, torture, and torment threaten to overtake us if we do not rally together to fight against it. In the twentieth century, for example, the list of prominent villains is so illustrious that there is no need to mention names. On a dramatic basis, there is no way to contemplate the events of the twentieth century without considering the infamous figures in world affairs. It was an intense era of villains and heroes both—the perfect fulfillment of this world's dramatic potential. The twenty-first century has already brought forth a new cast of villainous characters.

Enunciating the function that villains play in our world does not in any way mean minimizing the genuine pain that they can cause. In order for this dramatic reality to fulfill its design parameters, it is necessary that we experience a certain level of pain. The pain that was experienced by the victims of World War II, for instance, is impossible to quantify, yet that pain was not the result of a world "gone mad" but rather was the result of conscious design at the macro level. This earthly reality, I believe, is the prime art form of the macroconscious—it is the most compelling "fictional" world that it is possible to create. We who dwell here in this microreality help to meet the creative needs of existence itself.

People who assume the role of villains in this world do so because of their inherent energy characteristics and because of the guidance that they receive from the macro level. Following the guidance of the macroself will usually result in a significant sense of fulfillment, no matter how villainous are the actions that result from that guidance. Villains do not just happen. Villains behave the way they do because they feel empowered to act in the manner that is in accordance with their basic nature. A villain acts in a way that gives him satisfaction, and this satisfaction comes from the macro level.

It is always possible for the human self to resist the promptings of the macroself, because the human self is a matrix of energy in its own right. When an individual has the feelings related to a deep internal conflict, there may be a conflict involved between the macroself and the human self over a particular course of action. Although, in general terms, the macroself will have the "upper hand" in such conflicts, the human self may offer such a strong resistance to the designs of the macroself that the human self can prevail. Thus, a villain may choose to opt out of villainy, or a good person may turn to a life of crime. It should always be remembered that there is a certain degree of true complexity in the way our affairs play out.

The concepts of heaven and hell will be discussed later, but it should be mentioned here that the fate of a villain in the "next world" has nothing to do with religious portrayals of the afterlife. No one is consigned to "hell," no matter what his deeds may be. All theological thinking regarding rewards and punishments for our deeds in this world should be put aside. Macroreality does not operate on a basis that any religion comprehends. The nature of energy bears little relation to the moralistic notions of humanity.

Heaven, Hell, Karma

While heaven and hell can be considered mythical kingdoms, macroreality is not in any way a fantasy. For all of human history, religions have defined the afterlife using various fairytale conceptions that have no rational context at all. Gullible human beings have quite often bought into these portraits of the afterlife because they lacked any compelling alternative perspectives. The last several centuries have seen the rise of rational skepticism in regards to theological fantasies. However, skepticism that is taken to the extreme of existentialism serves no ultimate logical purpose.

When an individual in our earth microreality dies, its personal matrix of energy configurations will be attracted to a harmonious energy region of the greater reality—macroreality. This is an automatic process because like energy seeks out like energy. By the time of one's death, every individual will possess a unique matrix of personalized energy that serves as the true definition of that individual. Any being is defined by its actions, because each act serves as a creative direction that energy has taken. Even a small act will have some effect on the energy configurations of that being. Large acts will have a large effect. This is an unalterable process.

As an example, if someone steals something from a store at any point in his or her life, this act will forever be a part of that person's energy definition. No act can ever be "erased" or redeemed. An act of thievery will eternally define an individual as a thief, even if no other acts of theft ever take place. This does not mean that a criminal act—any criminal act—consigns an individual to eternal damnation, because it does not. It does mean that this action becomes an aspect of that person's energy definition, and there is no way to make it go away.

It should be kept in mind that no single act has the absolute power to define a person's personal energy matrix, regardless of what that act may be. Everyone's energy matrix represents the entirety of all actions taken during the course of one's lifetime. Acts of "evil" can be balanced out by acts of "good." Most people, throughout their lifetimes, will engage in a very wide variety of actions and the complex nature of all these acts taken together will be reflected in their energy configurations.

In some cases, of course, there will be no real balance of actions because certain behavior traits have dominated throughout an individual's life and will thus overwhelmingly identify that person's matrix. This world is indeed capable of producing both saints and monsters, people who have lived their lives in a strictly defined manner. A saint, upon death, will gravitate toward the energy region of macroreality that is harmonious with its energy configurations, just the same as a monster will. These places are not heaven and hell as portrayed by religions, but they certainly can be thought of as being quite distinct from each other.

While it is one's acts that have the most impact on the energy matrix, it seems reasonable to conclude that a person's ingrained thought patterns will have an impact as well. Thinking a particular thought repeatedly, fervently, over and over can represent the equivalent of committing a deed. A fleeting thought, no matter what the thought might be, should be considered to have no consequence whatsoever, but an obsessively repeated pattern of thinking will become an intrinsic part of any person's personalized energy. Thus, we should try to maintain an awareness of what impact our thoughts may have on our very energy essence.

We should see macroreality as being an infinitely complex universe of energy that accommodates unlimited variations of configurations of energy. When death releases us from our focus on this microreality, we will simply follow the strong attraction that we feel toward the region of macroreality that has an energy correspondence to our own matrix of energy. Because this region of macroreality is in harmony with who we are at our most basic level of being, it will feel quite "heavenly" to us. We will find ourselves delightfully truly compatible with the other beings who dwell there. This region of macroreality will be our eternal "home" because it shares our own energy definition.

It has been said that one person's heaven is another person's hell, and this does indeed have some truth to it. A region of macroreality that is harmonious to our energy configurations is heaven while the region that is utterly antagonistic to our energy patterns is undoubtedly hellish. However, the imagery of heaven and hell that religions have offered up over the past few thousands of years have no reality to them at all. In your heaven, there will be no angels strumming their harps, unless this is a personal reality that you wish to manifest for yourself, as in the case of a dream reality.

In hell, there are no fires of eternal damnation to be experienced unless, again, this is a personally created reality that, for psychological reasons, you wish to have the experience of. Hell will generally feel perfectly fine to those who find themselves there. The beings around them will be highly compatible and sympathetic in nature. There will be no sense of punishment unless punishment is something that they feel the need to endure. No mythical God figure has judged them. They are simply who they are.

It seems likely that there are many similarities between the "landscape" of macroreality and the environment that is found in the dreaming reality. In macroreality, the world that we perceive around us will suit our desires, whatever they may be, from moment to moment. It is a world of infinite creative manipulation, comparable to what we can experience in "conscious dreaming." We can live in the grandest of cities, with streets paved in gold, or we can dwell in a place of the utmost isolation—on the top of a mountain or in the midst of a desert. The world around us will be as stable, or as mutable, as we wish it to be.

Because it is the intrinsic nature of energy to be in a constant state of evolution, those who dwell in macroreality will continue their incessant evolution in a similar way to what was the case in this earthly microreality. We will continue to make an endless number of creative choices that will continue to have an impact on our personal energy matrix. Whoever we were at the time of our death is not who we are forever destined to be.

Even someone who was fully dominated by evil in this microreality can have the potential to evolve in new and different directions. The alternative is also true, that an individual who lived a saintly existence in this world may make creative choices that will cause him or her to evolve in possibly unexpected ways. Since there is no judgment made on our deeds here in this earthly reality, there remains absolute freedom to continue to evolve. It is human nature to want to see evildoers punished for their sins, but reality does not operate on those terms. Both the evil and the good fulfilled the roles that they were given to play in this dramatic reality, and thus the concept of punishment and reward is simply not applicable.

Given these considerations, it can be seen that the traditional concept of karma that has been a part of Eastern philosophy has no real validity. Notions relating to karma can seem quite reasonable to human sensibilities, and this is why the concept of karma has been so appealing to so many people for such a long time.

This earth reality is not a place to seek out or to achieve any manner of spiritual perfection or fulfillment. The thrust of existence, utilizing the macrocreation force, is the complete expression of the creative urge. Perfection, or spiritual accomplishment, do not relate to this intrinsic creative thrust. At every point of their existence, people are as "perfect" as they are capable of being. We are all absolutely "perfect" right now— right this very minute. There is no path to walk in life except the path of continual creative evolution. Karma simply has no reality at all.

It should always be understood that the full nature of macroreality exceeds the human ability to completely grasp it. All that can be offered here is a minimal introduction to this subject, using a basic framework of ideas. It is reasonable to conclude that the macroreality environment is capable of absolute duplication of the earthly environment, just as is the case with our dreams. Yet macroreality can also offer realms that are utterly abstract in nature and that bear no relation to anything that can be conceived by the human imagination. What is most clear is that what we will find in macroreality will bear very little resemblance to the fairytales that we have been offered regarding the afterlife.

In macroreality, it is presumably possible for a human being to evolve in a creative direction that will take them far away from all aspects of humanity, to cease being human at all. It may be possible to go even further, to evolve to a point where a being may very nearly cease to be— a close approximation of existential nonexistence. Most likely, these are options that we have if we choose to embrace them. In this microreality, we have been enveloped by limitations of every sort. In macroreality, there are no limitations on our potentials for creative evolution. This means that human beings can evolve into various nonhuman states of being if that is their choice for their ultimate development.

Of course, all beings who found existence in this earthly reality of ours will also find existence in macroreality. For human beings to believe that the afterlife is their sole domain represents the ultimate in irrational vainglory. In macroreality, human beings finally lose the self-importance that so defined them in this world and so they can finally appreciate the creative contributions that nonhuman forms of life make to the totality of existence. Our animal friends, as well as insects, plants, microbes, and everything else nonhuman in this world, will continue on their own path of energy evolution in the next world. Their futures are just as unlimited as ours are.

Macroreality is the "true" reality while our microreality can be thought of as a fictive reality. While here, we cling to this world as if it were the only one, but it is only one of many—an infinity of realities exist for the sake of creative fulfillment. The complete expression of our energy uniqueness can only be experienced there.

Reincarnation

The macrocreation philosophy contends there is little logic to be found in traditional ideas regarding reincarnation. If an individual is understood as constituting a unique set of energy configurations, then it would not be possible to combine one matrix of personalized energy with another one, which is what the idea of reincarnation amounts to. If one person can live many lives, then this would mean that the uniqueness of one person can somehow be merged with the pure originality of other people. Reincarnation has to be seen as something that is quite dubious on rational terms.

In the previous chapter, the unlikelihood of karmic principles was discussed. If there is absolutely no reason why more than one life in this earthly reality needs to be experienced, then there is no reason for reincarnation to exist as a corresponding aspect of karmic principles. We who dwell here in this reality do so to fulfill the creative needs of the macroconscious, and not because we are involved in the highly unlikely quest for some sort of spiritual perfection. The macrocreation force is not a spiritual force; it is a creative force that underlies the basic nature of existence. In contrast, the concept of spirituality is a human concept that often pleases human sensibilities.

However, none of this means that the numerous accounts that have arisen over the years of people experiencing some sort of "reincarnational memory" should be immediately derided as delusional. Many of those who have had these memories are the types of people who have been considered fully rational in the ordinary course of their lives, and what they have to report should be taken seriously. Nonetheless, it is quite reasonable to reinterpret these experiences in a different way that transcends traditional thinking regarding the concept of reincarnation.

In the most intriguing cases, the "memories" that rise to the surface of consciousness will occur in a young child—someone who has had absolutely no exposure to the culture or the era that is being recalled. Occasionally, there may even be verifiable tidbits of information that come forward, including such hard facts as names, dates, and specific place descriptions. If traditional ideas regarding reincarnation can be considered dubious on intellectual grounds, then an alternative explanation must be offered.

As is true with all things, the macrocreation answer lies with a fuller understanding of the nature of energy and the expression of our beings. First, it is likely that the sense of complete individuality that we experience in this microreality is not something that extends to our existence in macroreality. In ultimate terms, energy needs to be thought of as a continuum—there can be no way to achieve absolutely hard and fast boundaries within this continuum. While we dwell in this earthly reality, we are cut off from any real overt sense of this continuum (although some of us do enjoy the very real sense of being connected to all things), and so we live out our lives with the designed delusion that we are utterly individual in all regards.

In macroreality, however, we have a much greater awareness of our particular position in the energy continuum. We know that we exist as a focus of personalized energy, connected to the continuum but maintaining a sense of individualized perceptions as well. We understand that we are individual beings in a relative form, not in an absolute form. Each being of focused personalized energy is intimately linked to other beings who are composed of highly harmonic energy configurations. The beings with whom we share this quite intimate link can be considered our "relatives," though our connection to them is more multifaceted than the bond that we enjoy with our human relatives in this microreality.

One term that can be used to describe this harmonic-energy being with whom we share such a strong link is the "associated self." An "associated self" is a fully unique being, like we are, but it is a being that, to a certain extent, has "boundaries" that can merge with our own. In essence, this means that their lives are "open" to us, as we are to them. Some of the energy that constitutes our very self is shared with another.

In this world, there are people who, due to their very sensitive nature, can maintain a certain link with those beings who are their associated selves. Because, until now, there has been no proper framework to understand this link that can occasionally manifest itself, fanciful ideas about reincarnation have taken hold in the public consciousness. The concept that we can live a series of lives here in this earthly reality is a much simpler one and a much more readily appealing one than is a theory concerning the nature of personalized energy. Those memories that surface of other lives that have been lived are indeed valid, but not placed within the proper context.

For many people, the idea that we can return to this world countless times over the course of thousands of years is truly quite exciting and has great psychological appeal. In contrast, ideas relating to an energy continuum lack any sort of "romance" and thus may not compete terribly well with traditional thinking. But it is always important to remember that the universe is not human and truth is not human. What appeals to a human being will sometimes be simply something that comforts us and seems to make some kind of sense. We should always be willing to look deeper and to go beyond our tender human sensibilities.

The relationship between a self and an associated self is much, much closer than that experienced between human twins. To some extent, the boundaries that separate these selves are blurred and can be considered permeable. A being can only be thought of as a focus of energy, and where that focus fades, another focus takes shape and form. In this microreality, we can have the experience of nearly complete isolation while, in macroreality, that is not the case. The "reincarnational" memories that can float to the surface may feel like our very own, but the nature of our existence in macroreality reveals what is really going on.

The Angelic and the Demonic

This chapter will present ideas on how the energy morality spectrum may function in macroreality. While religions and New Age philosophies have presented their human-centered notions regarding angels and demons, the macrocreation concept of such beings will be very different. There is no reason why religious and New Age thinkers should consider this subject to be their own special domain.

Most human beings, while dwelling in this earthly reality, contain a moderate blend of "good" and "evil" qualities that come from the energy morality spectrum. In our world, there are relatively few people who can be considered the saintliest of saints or the most vile of evildoers. Most of us manifest a complex configuration of these moral energy aspects that prevent us from going to extremes of behavior. However, in macroreality, it is possible for the beings who are native to that reality to manifest extreme aspects of either "good" or "evil" energies. We can refer to these beings with terms from popular usage—angels and demons.

In macroreality, a focus of individualized energy that we can label a "being" may be constituted of concentrated energy configurations that come from the "evil" region of the energy morality spectrum, and this is why demons can have an actual existence that is completely apart from the role that they have played in popular fiction. It should definitely be understood that these demonic beings have nothing at all to do with theological conceptions—demons have a reality that no theology comprehends. Religions that recognize the existence of demons cannot be counted on to have a true and rational understanding of what demons actually are.

Having said this, there is no question that demons can respond to the efforts of exorcists and other religious workers who make it their solemn duty to "cast out" these evil beings from our world. Because demons represent an extreme configuration of energy, it may be that they are, in effect, "unbalanced" and thus not entirely cognizant of the particulars of their own reality. Thus, they would find themselves vulnerable to human will and human manipulation. Since a human being possesses his own independent focus of power, it is conceivable that a demonic being may find itself subject to a human's vigorous efforts.

A demon, not understanding who or what it actually is, may adopt the identity that is given to it by a human who has engaged it. In addition, the demonic being will have a much more vivid and creatively meaningful earthly experience if it can successfully play the role that has been assigned to it. From anecdotal accounts, it appears that these beings derive much meaning from the spirited interaction that they have with humans. If the demons were to announce that they actually had a completely different existence from what humans assume, then this revelation might spoil their "fun."

One question that should be asked relates to the reason that demons, and angels, have existence at all. Since their function and purpose has no theological foundation, there needs to be an alternative foundation. In the context of the macrocreation philosophy, beings who represent extreme energy configurations, such as angels and demons, owe their existence to the creative needs of the macrocreation force. These extreme beings manifest a particular configuration of energy that the macroconscious must find creatively quite satisfying and necessary. The macroconscious, operating from its nonhuman sensibility, can appreciate the "charms" of the demonic in a way that a human being could not.

Over the years, people who consider themselves the embodiment of rational thinking have tended to dismiss the concept of demons due to the theological context in which demons were placed. The macrocreation philosophy offers a way to look at the subject of "supernatural" beings that has no relation to theologies or New Age thought. Because of this, a rationalist who clings to his old intellectual framework primarily because of the comfort level that it brings him cannot really claim to be a rationalist at all.

Throughout history, there have been fascinating reports of demonic "possession" cases that have been terrifying in their implications, if factual. In some instances, these cases have been well-documented by fairly objective observers and thus have the aura of authenticity. While the religious aspects of these incidents are certainly open to dispute by the rationalist skeptic, the macrocreation view is that it may, in fact, be possible for a demonic being to "possess" a human being.

However, it seems reasonable to conclude, using the macrocreation framework, that an incident of "possession' can only occur if there is cooperation of the victim and the victimizer at the macro level. Due to the extremely high dramatic potential of this sort of demonic engagement, the macroself of the victim will serve as a willing accomplice. The macroself, in its quest for a creatively satisfying course to follow, will, upon occasion, lead the human self into situations that may involve intense pain and suffering.

The fact that cases of "possession" are reported quite infrequently demonstrate that this type of occurrence is something that rarely gains the approval of the human self at the macro level. It would not be beneficial for this world of ours to appear to be infested with demonic activity. Demons, like ghosts and UFOs, tend to be subjects that, for most, skirt the edge of human awareness. For this reason, very few people will have encounters with the demonic—only those individuals who are sufficiently courageous at the macro level to take on this tortuous experience. Though the exorcist will frame this encounter as a battle between "good and evil," it is actually a creative affair that is playing itself out on this earthly stage.

While humans tend to be understandably in awe of the "supernatural" world, it is possible to gain the kind of perspective that will allow a person to deal with these very frightening things with a greater degree of equanimity. Demons, and some ghosts, truly enjoy scaring people simply for their own amusement in a way that is very much the same as one human being scaring another. As a general rule, their essential power is a psychological power—their will is imposed on us in order to accomplish their creative goals.

It helps to understand that the term "supernatural" really does not describe anything at all. All beings in existence have the same fundamental nature—a focused matrix of individualized energy. The primary difference between a demonic being and a human being is that the demonic being is focused primarily in macroreality while a human being is focused primarily in this microreality. Of course, a demon is "evil," but there are some evil humans as well.

The primary leverage that demons have over humans is the fact that most humans have absolutely no comprehension of what they are dealing with. If they imagine that they do have knowledge about the demonic world, that knowledge will be saturated with theological mythology and thus may make the whole situation even more frightening. A human being who collapses in fear in the face of a demonic attack has lost the battle at the outset. To surrender to one's fear is to become absolutely defenseless.

A human being possesses sufficient personal power to make a "good fight" out of any encounter with the scary supernatural. Because a human is a being of pure energy, we are never entirely helpless, no matter what we are facing. We can summon the force of our personal power by focusing our intent. Our intent can be described as our energy being manifested as an idea or a goal or an emotion. When we summon our full intent, we become as focused as a laser beam.

All of us possess at least a modicum of "dark" energy aspects. We can use this dark energy to become rather scary ourselves—a fearsome being in our own right. We have within us what we need to fight back in a demonic attack. However, the opponent must always be respected for the power that it manifests, and we should do all that we can to avoid unnecessary involvement with the demonic world.

Besides demonic beings, the energy morality spectrum also offers their opposites, those beings who are known to us as angels. Religions make the claim that angelic beings serve as representatives of their particular religion and their particular god, but this claim has no real basis in fact. However, in many cases, angels will indeed take on the identity that humans with whom they are interacting wish to give them because this will greatly facilitate the ease of interaction. An angelic being will have very little motivation in trying to explain the actual basis of its existence to a person of religious faith. It will generally serve the angel's purpose to take on the role that a human will readily accept.

An angelic being will come to the assistance of a human being if that human's macroself believes this intervention will serve its best interests. Thus, an individual of very strong religious beliefs is much more likely to receive a visit from an angel than is an atheist. For an atheist, this "supernatural" assistance may be a cause of disruption and disturbance in their lives because it goes directly against their deeply held beliefs. Therefore, assistance that is provided to someone with no religious beliefs may be in a form that will not be threatening to them—it will be camouflaged in such a way that its "supernatural" origins will be hidden from view.

Angels, despite their demeanor of benevolence, are as fully capable of telling lies as are their demonic counterparts. Our human standards for "truth" and "falsehood" simply have little or no relevance to a macrobeing. Since macrobeings understand the fundamental creative flow that informs our existence, they are much more tuned in to this ultimate creative expression than they are to human moralistic concepts. If it serves a valid creative purpose for an angel to lie to a human, then the angel will have no hesitation in doing so. There may have been countless incidents where, for example, an angel may have identified itself as "Michael," or as some other recognizable figure, simply because it knew that this identity would have a very favorable impression on the human with whom it was interacting. The angelic being in these cases is not being devious; it is being creative.

A falsehood of this type has no malicious intent, but there is no question that a human is being misled. The human concept of "absolute truth" has no meaning in the realm of macroreality. The beings who dwell there operate on their own motivations, and these motivations may not always be comprehensible to the human mind. Religions make the claim of understanding "heaven" and those who dwell there, but religions, in almost all cases, have no real information concerning heaven and what they have to offer is myth.

As long as religions play the major role in society that they have been playing for thousands of years, the occasional appearance of an "angel sent by God" will still occur. Angelic beings will work with whatever belief system humans employ—they have no desire to transform human thinking. Angels come into our lives when their presence is needed. They do good deeds, and they perform good works because their matrices of energy configurations come from that area of the energy morality spectrum that we identify with "goodness." Angels are indeed good, but they do not come from God.

In contemplating the beings of macroreality, ranging from the angelic to the demonic, one other macrobeing of myth and legend can be considered in terms of energy—Satan.

The macrocreation position is that, in terms of an independent entity, there is no such thing as God and there is no such thing as

Satan. What has been theorized as the macroconscious is something that accommodates the entirety of existence—it is the awareness that accompanies the macrocreation force. Therefore, the macroconscious can be said to manifest all the attributes that have been attributed to God as well as all those attributes that have been attached to Satan.

As stated previously, the macrocreation philosophy contends that there is no battle being waged between the forces of good and the forces of evil. Everything in existence is the manifestation of creative expression, including all those things that we label "good" and all those things that we label "evil." If this is in fact the case, it certainly does turn all theological belief systems on their heads. No one is battling for our soul. No war is being fought. No sides have been drawn. Nothing is black and white.

Ghosts

Rationalists understandably have some difficulty in accepting the reality of ghosts, because encounters with these entities can certainly be of a highly subjective nature. When there is just a single person in a room who has taken notice of a "ghostly presence," it is quite easy for the others in the room to dismiss the incident. While a great number of ghostly encounters can be disregarded, not all of them can be, and the subject deserves to be taken seriously. The reality of ghosts has been considered a dubious prospect primarily because self-anointed rationalists have made it a part of their dogma. This idle dogma needs to be re-examined.

The individualized energy that comprises a being is an intrinsic aspect of the energy continuum—this is true for ghosts as well as humans. No being exists entirely in our microreality or entirely in macroreality. A human being has its primary focus in this microreality, but we extend, as part of the energy continuum, into macroreality. For a ghost, it is the reverse. A ghost is a macrobeing who has chosen to manifest part of its individualized energy in our reality, for its own particular purposes. Our world has maintained a strong attraction for them, and so they opt to retain some of their energy and personality here.

From the many anecdotal reports that have surfaced over the years, it seems clear that beings who are focused in macroreality continue to take a fairly strong interest in the affairs of this microreality. Our world exists as an arena for manipulation on the macro level and, thus, those who were once human will still find this world fascinating, especially because they now have an entirely fresh view of how the affairs of our world are managed. We humans are apparently always in the company of others, those beings from the macro level of existence, without being at all aware of it. When we do become aware of it, then we are seeing "dead people."

A macrobeing who focuses a sufficient degree of their individualized energy into our microreality will be able to make their presence known—as a ghost, apparition, spirit, poltergeist, or some such other manifestation. The motivation that a macrobeing will have for manifesting itself in our world will certainly vary from being to being. Often, strong emotions will be involved. Macrobeings can be as benevolent or brutal as humans. There is not much point in making broad generalizations.

In macrocreation philosophy, there is no such thing as a "soul." A soul has no definition in terms of energy. New Age thinkers have offered many ideas to explain the reason why ghosts walk among us. Most of these ideas can be attributed to human-centered notions of the afterlife, which have no particular correlation to the actual nature of what can be termed macroreality. Ghosts can best be understood in terms of energy and in terms of personality. They need not be considered especially mysterious at all.

A ghost engaged in a haunting of a particular locale can be thought of as something that is akin to a "split" personality. Most of the energy that comprises their complete personality remains focused in macroreality, while a splinter of that energy has found its focus here. Since a ghost is a mere sliver of personalized energy, its accompanying personality is likely to be fairly simplistic and easy to comprehend. Thus, those people who deal with ghosts who inform the entity that all they need to do is "go to the light" will often have favorable results. The minimalistic nature of a ghost's personality is often easy to manipulate.

A ghostly being may have very little self-awareness and precious little understanding of its own reality. It finds itself in our reality in the same way as we may find ourselves in a very strange dream reality. There is a lot of confusion, and we are generally quite glad to wake up in a reality that is more familiar to us. The ghost will also, quite often, be happy to "wake up" as well.

However, there will be some ghostly entities who retain a fairly large measure of their personalized energy while focused in this microreality. These beings will offer a much greater challenge in attempting to deal with a haunting situation of their creation. These macrobeings have a strong motivation for being here, and this strong motivation allows them to summon a higher level of their individualized energy. Enough of their personality may be focused in this world that they will have awareness of who and what they are. They will not be confused and they are not interested in "waking up" back in their beds in macroreality. They have a mission to fulfill, and they intend to see it through.

All ghostly beings need to be treated as firmly and as forthrightly as possible. Ideally, a human interacting with a ghost should gain full control of his or her personal fear. In some cases, a ghost may choose to present itself in a form that is meant to be quite frightening, not much different from a human putting on a scary costume. Thus, we should treat a ghost who is attempting to frighten us in the same manner as a human being who is attempting to scare us—we should not yield to an unreasoning fear. Ghosts have a reality that is different from ours, yet not so very different. We are all beings of energy.

Channeled Information

Too many people in the New Age community have put their faith and trust in "channeled" information that comes from a self-proclaimed source of higher wisdom. It is quite naive for anyone to assume that a personality who is focused on the "other side" is automatically wise, knowledgeable, and fully trustworthy simply because they are where they are. It should be understood that beings whose focus is on macroreality can be just as devious, deluded, confused, grandiose, and self-serving as any human being might be. Macrobeings are not automatically impressive characters.

In fact, those macrobeings who seek out human contact in order that they may pass along their "wisdom" should be considered suspect right from the very beginning. A being who dwells in macroreality has sufficient knowledge and awareness to realize the extreme difficulty in trying to educate human beings about the actual nature of reality. In general, it is the macrobeing's desire to inspire awe and veneration in the human contact that motivates them to "infiltrate" our world. They are not, as a rule, truly interested in bringing enlightenment to humanity.

A macrobeing who has humanity's best interests at heart would be quite hesitant to offer information that is quite inconsistent with most people's belief systems. A human who is the recipient of this channeled information may very possibly have his or her world turned upside down, causing great psychological stress. The human mind generally has a perspective on life that is so far removed from the macro perspective that any attempt to bridge that gap is a very risky endeavor.

In contrast, an unenlightened macrobeing, seeking its own self-glorification, will have no particular concern with putting its human contact at risk. Once this link has been established, the macrobeing will deliver tidbits of "wisdom" that will usually fit in quite well with the recipient's overall philosophical framework. While some aspects of this knowledge may seem new and fresh to the recipient, it will nonetheless seem quite reasonable and believable to them because it fits their own personality structure. The being who is imparting these shards of enlightenment basically just wants approval from its human partner, so it will tailor its message accordingly.

One example of this is the channeler who receives truly wonderful news about heaven—about how its streets are paved in gold and where all residents every minute of their glorious existence devote themselves to paying eternal tribute to the marvelousness of God. A scenario such as this is, quite obviously, a fairytale designed by an eager-to-please being in macroreality. Many people form their ideas regarding the afterlife when they are small children, and these concepts can retain their emotional appeal into adulthood.

When dealing with the subject of channeling, it needs to be understood just who or what is the source of the channeled information. When the human being involved in this process goes into a trance, there is the distinct possibility that nothing more than a disassociation of personality is being achieved and thus no "independent" entity in macroreality is involved. The complex nature of the human personality makes it difficult to be certain whether the macro-level entity who is involved in the process exists "inside" the macroself of the human or "outside" of it.

It may be that the macro personality can be characterized as an "associated self," similar to what was discussed in the case of reincarnational personalities. If this is true, then the channeled entity could be considered to be both "inside" and "outside" the macroself. The complex nature of the energy continuum as it relates to human personality makes it challenging to sort out. Regardless of this, it should be emphasized that there is no reason whatsoever to pay special heed to information that comes to a human being from the macro level of reality. All information needs to be subject to rational analysis, regardless of its source.

One important fact to keep in mind is that every one of us is the recipient of "channeled" information every single day of our lives. This is a quite natural process and is usually completely effortless. Any insight that we receive from our macroself can be considered channeled in that this guidance originated from the macro level of reality. We receive these insights quite routinely, so routinely that we often pay little attention to them. The "channel" from the macroself to the human self is always open, and it is always available to us. No special effort on our part is necessary to receive this guidance.

When the process of channeling becomes a forced and "fancy" affair, this is when trouble will loom. The classic channeling scenario is when an individual goes into a trance, thereby submerging their usual human personality, and what emerges is then considered the "trance personality." Quite often, this new personality will speak with a vague foreign accent or have some other exotic cultural attributes. There is a psychological need on the part of the trance medium that the "higher being" who speaks through him or her must have personal characteristics that are, in their own way, rather impressive and interesting. This sort of channeling endeavor is essentially a mind game. Genuine enlightenment is not likely to result from these fanciful methods.

Wisdom and enlightenment are things that simply are not available to us "on demand." If we manage to become a little bit wiser as we live our lives, it is because we have kept ourselves open to new ideas, insights, and interpretations as we struggle to achieve what we want and need. As a general rule, wisdom comes to us in dribs and drabs—tiny droplets that very gradually combine into a deep pool of utter clarity. This process takes patience, and it takes humility. We need to pursue wisdom without ever necessarily expecting to truly achieve it. The usual process of channeling involves a deluded human being linking up with a deluding personality who is focused in macroreality. Such a path as this will not lead to enlightenment.

Near-Death Experiences

The typical near-death experiences that have become the standard portrait for our existence in the afterlife should not be taken as a meaningful depiction of macroreality. One way to understand the common near-death scenario is to compare it to a visit to a far-off, exotic land—but visiting only the ultra-modern airport terminal. As world travelers know quite well, what is experienced at the airport does not necessarily provide much indication of the land that lies beyond. The standard vision of the afterlife that these encounters have to offer is meant to provide comfort to the fearful human mind, and that is all.

The transition to macroreality is a challenging one to make because most humans have no real intellectual preparation for the nonhuman nature of the universe. For those who have a strong religious background, or even a weak one, what they want the afterlife to be will be provided for them, courtesy of their own macroself. Every reality to be found in the vastness of macroreality is a created one, including these introductory experiences. The initial encounter with this new world is meant to soothe, not to shock. Nothing could be more soothing to most people than to be greeted by their departed loved ones.

Even those people who have little religious faith may still have much the same near-death experience as those who do. The near-death event is designed to be as comforting and reassuring as possible. To accomplish this, the macroself is perfectly willing to adopt the scenario that will do the best job for the situation, even if this will serve to strengthen the religious depictions of the afterlife. The full nature of macroreality will eventually be revealed to one and all, so the initial feel-good production will have no long-term implications.

Most people avoid thinking about death, especially their own. Because of this, the standard portrait of the afterlife that arises from the cultural milieu will generally be adopted by almost everyone as the easiest course to take. People have a very strong need for reassurance, and the standard near-death experience provides this. Even for the agnostic, the mere possibility that upon death, we will be greeted by our departed relatives and surrounded by a warm, beautiful white light that represents pure love is very hard to turn away from.

Most of these near-death memories will be brief and simplistic. These moments of introduction to the afterlife involve a human being's most basic emotions and links to essential aspects of human psychology. It will be a long, gradual process of re-education—an individual's strongly held belief system cannot be shunted aside too quickly. As the comforting religious trappings are withdrawn from the initial presentation, the simplistic restrictions of religious belief systems will become increasingly apparent. Most people will receive this new information reasonably easily and will make the necessary adjustments to their philosophical framework. For some others, the transition will be longer and will take much more effort on their parts.

It can be asked if, when someone who is newly arrived in "heaven" is greeted by deceased relatives, this experience is truly what it appears to be. Since the initial near-death reality is one that is created by the macroself in order to provide a reassuring transition to the afterlife, it can be debated whether the images of the deceased loved ones are as "real" as they might be. It is quite likely that not all of the reassuring images that surround a new arrival at this time are what they may seem—it is so simple a task for the macroself to create a friendly environment that there would be no real need for possibly dozens of family members to "make the trip" for this occasion. Some of those who come to greet us may be "real," while others may not be.

In macroreality, a being who was once part of this earthly reality may have chosen creative paths that will have taken them very far away, in terms of individualized energy, from who they once were. Thus, "Uncle Edgar" may have been a fond figure in someone's earthly existence, but he has now evolved into a particular being whom the new arrival would barely be able to recognize. In addition, the being who was once kindly old "Uncle Edgar" may no longer have much interest in the arrival of his long-ago niece. If "Uncle Edgar" were to show up at all at this event, it would presumably be in just a minimal fashion, presumably sending a small sliver of his personalized energy to the arrival ceremony. Whatever the case may be, it is best to avoid simplistic, feel-good assumptions about these near-death events. Macroreality is infinitely complex and intrinsically nonhuman, so there will always be many possibilities to consider.

For most people, their conception of heaven was crafted in their early childhood years, and these images became ones that held so much emotional importance that they could seldom be strongly modified, let alone completely disregarded. In many ways, adults are actually merely large, old children when it comes to their basic belief systems. As we age, we will mature in many ways, but intellectual maturity will often be lacking. To abandon beliefs that provide us with a great degree of comfort and reassurance is extremely difficult to do, and it is not something that one person should ever ask of another. It is not "wrong" to remain a child all of one's life, but it certainly is a severe intellectual restriction.

The typical near-death experience that has been brought back to us by the survivors of these events should be thought of as being more "real" than "not real." These experiences are designed to ease our transition into a new and infinitely more complex world. The average person will find these typical "heavenly" events quite soothing and wonderful, and this is what matters most. Despite the lack of anecdotal reports, it seems highly likely that there is a wealth of possibilities for us to experience in the initial moments of our afterlife existence, going far beyond the rather tepid scenarios that have been so commonly brought back. When you die, expect to be reassured upon your arrival in macroreality, but you should also expect some phantasmagoric surprises.

Death as a Creative Act

As a general rule, death is by design. Death occurs when, where, and how it best serves the creative maneuverings of the macroself. Due to the very real element of randomness that is at work in the world, however, there presumably still is the possibility that a death can be considered accidental in nature, though this would be a rarity. In almost all cases, the "angel of death" can be recognized as our very own self, functioning at the macro level of reality.

In most cultures, death is considered a tragic event, to a greater or lesser extent, depending on the circumstances. Even those who profess strong religious faith can have great difficulty in accepting "God's will" when the death of a loved one is involved. Possibly, at a deeper level of awareness, even the religious faithful realize that the portraits of heaven that they have embraced in the course of their personal philosophical development must, in the end, be thought of as merely fairytales. Thus, when a death occurs, the bereaved individual will experience a profound conflict between the concept of heaven that they want to believe and the general suspicion that those heavenly images cannot possibly be true.

Therefore, nearly everyone in this world, from the very faithful to the very atheistic, has had grave doubts regarding the continuation of life after physical death. For most, portrayals of the afterlife have been so vague and so very dubious that they have held very little resonance. Emotionally and intellectually, death seemed to be the end of existence. Death came to be seen as a thief, a scourge, a destroyer. In our collective imagination, death has loomed menacingly for all of human history.

But it is possible to see death in a very different light. This microreality is a realm of extreme conditions—pain, frustration, and trauma can be found in agonizing abundance. This earth reality is a tour of duty that we must endure so that the creative demands of the macrocreation force can be satisfied. While there is no question that this world offers pleasure as well as pain, the fundamental nature of this reality is to serve as an arena of challenges, not as an oasis of delights. Death, when it occurs, releases us from any further obligations to meet the requirements of this reality. Death is our liberation.

The earth reality can be seen as a lesser reality, in every way, to the central reality of existence—macroreality. The human self is a lesser self, in every way, than the vastly complex macroself. The system of linear time that we endure here is a monotonous grind compared to the "eternal present" of nonlinear macrotime. When the time for death comes, it means that our servitude has been completed and our freedom awaits.

In our society, it has been commonly noted that prisoners who have served much of their lives behind bars will experience significant anxiety as the time for their release approaches. Many prisoners have learned to embrace the strict boundaries and limitations that prison life has to offer. After awhile, prison has become "home." For some people, it is easier to live without freedom than to live with it. Our fear of what will happen to us at the time of our deaths can be compared to this.

While it is a good thing to not fear death and to yearn for the greater possibilities that macroreality has to offer, the premature closing of one's earthly existence by the act of suicide is not something that should be contemplated except in extreme circumstances. Despite what religions have to say on the subject, there are no moral implications to committing the act of suicide. However, the choice of suicide is one that does have quite significant implications for one's future existence in macroreality.

Every act that we commit in our lives, from the smallest to the largest, has consequences for our personal energy matrix. Our actions literally define us. If suicide occurs as the result of immaturity, self-absorption, self-importance, and self-indulgence, then this action will be sharply regretted on the "other side." Even in "heaven," emotional pain can still be felt.

Because this earth reality can be thought of as a dramatic reality, each life that is lived here is the story of a heroic protagonist and the adventures that he or she experiences. A life story that ends with a rash and pointless act of suicide will be profoundly lamented because it brings such a highly unsatisfactory conclusion to this particular story. While there is no sanction of "sin" associated with an impetuous act of suicide, there will be severe self-condemnation. This overwhelming sense of regret will be punishment enough.

This does not mean, however, that all acts of suicide will lead to self-condemnation and regret. A self-inflicted death that comes after a long and valiant struggle with intense pain, whether physical or emotional pain, will not instill a strong sense of guilt or regret. An individual who has made the struggle, year after year, to meet the challenges that life has to offer but who, ultimately, is not quite able to overcome these challenges should not be ashamed to commit the act that will finally end what has become unbearable suffering. Every act of suicide or potential suicide needs to be fully examined on its own merits. There are no "rules" regarding this. There is absolutely no one to pass judgment on what you do.

The macroself will generally be a partner in the decision to take this final step. The macroself may even facilitate the process if it has no creative objections to it. Most often, an act of suicide will be something that is strongly desired by the human self, and the human self will provide the full impetus for this act, while the macroself simply "stands back" and does nothing to thwart it. As a creative act, seen from the macro perspective, suicide can be considered either entirely valid or entirely invalid, depending on the unique set of circumstances involved.

A lovesick teenager who kills himself or herself due to a broken relationship is perhaps the most acute example of a suicide that will be very deeply regretted in macroreality. Because the human self is a power source of its own, even the most rash acts of suicide can still take place despite the strong objections of the macroself. If we are determined to do something "stupid," even if it leads to the end of our lives here in this microreality, the macroself cannot be relied upon to stop us. In many instances, the macroself will be able to engage in various maneuverings that will help to thwart some of our stupid actions—a suicide that could have occurred to someone at age sixteen may be outmaneuvered at that time but not later, at age eighteen, for instance. The creative interplay between the human self and the macroself is far too complex to paint a simple picture and achieve a simplistic understanding. The one thing that can be said with certainty is that suicide is an act with vast consequences.

When contemplating the subject of death, the possibility of the occurrence of an accidental death should be considered. As a matter of logic, the idea that a random factor is in operation in our lives has to be embraced. On a daily basis, so many truly small and trivial events occur so frequently that it would be quite unreasonable to imagine that these events were all the result of creative design at the macro level. Therefore, the concept of the random factor is quite indisputable. The central issue then is to attempt to make a determination as to how this factor of randomness actually functions and affects our lives.

As an example, if, on a particular day, we opt to have a tuna sandwich for lunch, this choice is likely to be made from a basis of pure randomness. However, if the tuna in that sandwich has become tainted and thus makes us ill, then it may not have been a random event at all to have had tuna that day. If the resulting illness is a very mild, then randomness may still have played a part. A severe illness presumably would only occur as a result of creative design. A choice for a lunchtime meal can bring forth a full consideration of the factor of randomness.

It should also be emphasized that an illness of this sort could occur because it is the design of the macroself that it should occur, for creative reasons. Also, in this scenario, the macroself may not have in its designs an illness for the human self at a particular time, but may choose to embrace the random factor at this time and allow the illness, due to its mild nature, to take place as a creative choice of its own. At the macro level of design, the factor of randomness may be both embraced and rejected on a frequent basis. This fact would make the process of creative maneuvering interesting and satisfying for the macroself.

It is not reasonable to assume that there could be an arbitrary cutoff point for this random factor to be at work in our lives. In order for the random factor to be fully functional, it must be able to be in force at every level of action in this earthly reality. The factor of randomness is a creative, independent force, and the macroself must be able to both utilize it as needed and to maneuver around it as well. Perhaps the macroself does not possess the ultimate ability to outmaneuver the random factor in every possible set of circumstances and, therefore, the random factor must be yielded to at certain times. Death of the human self by accidental actions may be the result.

A death that comes about as a result of random acts may be embraced by the macroself because it was unable to prevent it or because it chose to accept this death for creative reasons even if those reasons differed from its overall design plans. The macrocreation force needs to be understood as intrinsically unrestrained and unrestrainable. The death of a human being due to random, accidental factors can be embraced by the macroself so long as that death serves an ultimate creative need. The designs of the macroself may be pushed aside by the mighty creative flow of the macrocreation force in service of the infinite creative maneuverings of the macroconscious itself.

In one possible scenario, the death of an elderly patient in a nursing home may have been designed by the macroself to be slow, peaceful, and uneventful. However, an incompetent nurse may, at a particular moment, administer an incorrect dosage of medicine that may serve to hasten that person's death. Though the macroself may not have designed that death, it may nonetheless yield to the factors that led up to that death. The insatiable creative flow of the universe cannot be denied.

Then, there are acts of murder to be considered. Crime, in all its aspects, has a strong creative role to play in this microreality. Violent criminal acts can occur in this reality as purposeful creative acts designed by the macroself to impact our lives in the powerful way that they do. Murder itself needs to be seen for what it is—a design feature of our world. We can die at the hands of another because the macroself has made this act part of its design for us or because it allowed it to happen as a result of accepting the factor of randomness as it plays itself out.

Most likely, there is no crime and violence, as such, to be found in macroreality, at least not in the same manner that it manifests in our microreality. In this world, we live amidst the criminal element—they may be our next-door neighbors or members of our own families. However, at the macro level of existence, where like energy seeks out like energy, those beings with certain "evil" tendencies will tend to be found in the same energy region. This fact makes that type of energy easy to avoid for those of differing energies. Our world here throws us all together, which is one reason why this microreality needed to come into existence, to accomplish what could not be done otherwise. In this reality, we walk among murderers.

It should also be remembered that the moral systems that we employ here in this world do not have any automatic echo in macroreality. Even the act of murder is not a case for judgment in macroreality. All acts are understood to be, ultimately, creative acts that have creative repercussions, rather than moral ones. These creative energy consequences are, in their own way, as "severe" as any judgment laid down by a moralistic "God."

In order for an act of murder to occur, it is likely that there must be cooperation at the macro level between the selves of the victim and the perpetrator. The macroself of the victim will opt to utilize the dramatic choice of murder because this choice offers the most meaningful creative direction to go in. At the human level, of course, the choice of the macroself to be the victim of murder may be something that will be quite incomprehensible, but there is no reason why the creative maneuverings of the macroself should necessarily be understood by the human mind. The macroself has such a different perspective to the human self that to ponder the choices it makes that affect us here in our reality is much like a humble human being questioning the choices that "God" has made in his or her life. To borrow a familiar phrase, the macroself operates in mysterious ways.

For thousands of years, human beings have often wondered why "God" does the things that he does. We have wondered why "bad things happen to good people." No one ever succeeded in comprehending the mind of "God," although delusional individuals with strong religious beliefs imagined that they did. The macrocreation philosophy offers an explanation for acts of murder and other foul acts that may not be at all comforting or reassuring but at least fits in with an overall philosophical framework that is coherent and internally consistent. We may never acquire an emotional understanding of the murder of a loved one, but we can have intellectual insight into that act.

Of course, no matter how enlightened society may be regarding the subject of crime, a formal system of punishment needs to remain in place. Human beings will always require rules to live by, and there will always be the necessity of formal restraints against our occasional extreme behavior. However, if it becomes recognized that all crimes, including murder, must require cooperation of the victim at the macro level, then this acknowledgement should help to lessen society's tendency toward hysteria when it comes to properly handling violent crime. Rather than react with emotion-fueled rage, society may be able to respond to violent criminal acts with a rational degree of comprehension of the creative aspects that are involved.

Presumably, there will always be murderers in our world. An enlightened criminal justice system will combine adequate elements of punishment and confinement along with the full acceptance of the creative aspects that are involved. All traces of our Old Testament attitudes toward crime will, hopefully, someday be eradicated from society to be replaced by a more sophisticated awareness of the creative relationship between victim and victimizer.

Whether death comes from the hands of a murderer or in any other way, it should be seen for what it is—a creative act. Sometimes, this creative reason may be fairly readily apparent. Many times, we can observe that the deaths of a long-married couple will occur in close proximity to each other. Mainstream science can offer no particular explanation for this phenomenon, but most people can easily understand what is transpiring—one spouse has chosen to follow the other. This fact shows that death can be a choice that we can make. In most cases, the aspect of choice at the time of death is not readily apparent, but it is still there.

The creative ramifications of death are always quite significant. One person's death will always have an impact in this world—we are all players in a drama, and when one player departs the stage, the relationships of the other actors will change. A human being may not always be able to figure it out, but things happen for a reason, especially our deaths.

Illness and Injury

When illness or injury strikes us, or someone we care about, we should generally assume that a valid creative purpose is being served. It may sometimes be possible to deduce the creative impetus for this occurrence, and other times it will not. Either way, at the macro level of your existence, creative maneuvering is taking place that will have major consequences for you in your life.

In the most obvious cases, the occurrence of illness or injury will be simply to slow down the hectic pace of one's life. For too many people, their lives have a simple-minded goal to pursue that provides very little creative satisfaction at the macro level. Making money and acquiring prestige in a career are not unreasonable goals in and of themselves, but if this is all that is going on in a person's life, the creative rewards will usually be fairly minimal. There is no such thing as a workaholic macroself. Materialistic goals are usually simply tolerated by the macroself and not strongly encouraged. At some point, an illness or injury incident may need to occur to alter the one-dimensional nature of a person's life.

Human beings tend to fall into deeply ingrained patterns of behavior that can be difficult to redirect. In order to remove the human self from a rut that offers little satisfaction to the macroself, an occurrence of illness or injury can be arranged. Since most people resist devoting much of their precious free time to quiet reflection and contemplation, a shallow, one-dimensional personality structure often results. Thus, the human self may tend to bore the macroself. When this is the case, the macroself will use whatever creative means are at its disposal to change the creative dynamic of the human being's life. An incident of illness or injury will often serve this purpose quite well.

Once the illness or injury has occurred, the subject of pain will often come to the forefront. Usually, the level of pain that we will experience in these situations will be fairly tolerable. The pain or discomfort will be significant enough to get our attention and to affect our lives as much as it needs to in the service of creative design, but it will be a level of pain that we will be able to tolerate fairly well. Pain can enhance the dramatic quality of one's life, and this is often the creative role that it will serve. This suffering can even help us to feel more "alive" as we deal with the intensity that it brings to our daily existence.

However, upon occasion, the degree of pain that has been instigated at the macro level may be too much for us to bear at the human level. For its own creative purposes, the macroself may choose to push the dramatic potential of a pain situation to its very limits, beyond what may be acceptable to the human self. When this is the case, the human self must learn to set limits that are acceptable to it.

The level of pain that an individual experiences will relate to the amount of pain that that individual accepts in illness and injury situations. For example, when someone is experiencing a backache, there can be no defined degree of pain that necessarily must accompany that backache. The amount of pain that is felt in regards to the backache will be the degree of pain that is embraced by the human self as a creative aspect of that situation.

To put it another way, there is no medical reason why the pain that is experienced in that backache situation need have a particular degree of pain associated with it—the person who has the backache allows a certain amount of pain to be experienced, and this level of pain need not be as great as the macroself has designed. The amount of pain that will be experienced is a result of cooperation between the macroself and the human self, and the human self can learn to be a little less cooperative.

There is plenty of room for the human self to maneuver when it comes to pain—there is always something that can be done. The greater the level of pain that is inherent in the situation, the more possibility there will be for the human self to assert its own creative power and lessen it.

In order to reduce the degree of pain that the macroself has designed in a particular situation, the human self needs to focus its intent to a sharp point. This means that we must demand of the macroself that our pain lessen or even cease completely. This relationship between the human self and the macroself should not be compared to a humble child of God praying to a tyrannical deity to please take pity and do something nice for me. That is not how it works. Instead, a human being gathers together the personal power that is intrinsic to his energy matrix and uses that force as way to make a demand of the macroself. This force of unwavering intent will always have an impact on the macroself, and you will be listened to.

This does not mean that one can simply "wish" that pain goes away and never bothers you again. It is not always easy to summon the full force of your willpower in every single illness or injury situation. Your motivation in each situation has to be sufficiently strong. It is easiest to summon your will when the pain that you are experiencing is close to unendurable. When the stakes are that high, your motivation will be readily forthcoming. In some other situations, such as with fairly low-grade chronic pain, it can be much easier to simply "live with" the pain rather than fight it tooth and nail, so we tend to simply make the best of it that we can and do not assert our potential power.

Even in chronic pain situations, we can still focus our intent to some degree. One way is to choose, as a creative maneuver, to ignore the pain and act as if it was not there. This simple act can be quite effective. When we deny pain its reality in our life, then it can, in many instances, just fade away. Pain has to be accepted in order for it to be experienced. Pain plays a creative role in our lives, and so it must be acknowledged as something to be reckoned with, but the degree of pain that we allow in our lives is a factor that we always have participation in. Whenever the degree of pain that we are experiencing in any situation of illness or injury is too much to bear, we have the ability to lessen it. Upon occasion, when we have focused our intent truly well, we can banish pain completely, at least for awhile. The thing to remember is that we are never powerless—we are powerful and creative.

A creative aspect can also be found in the role that medications play in our illness and injury scenarios. The treatment that we utilize will prove effective or not at all effective depending upon the creative ramifications of the illness/injury event. All medications that are in use in our world can be considered no different than voodoo potions—medicines work or do not work depending on the creative maneuverings of the macroself that are being played out.

When we use an antibiotic, as an example, it may cure a particular condition, or it may not. Medicines, like all things in our world, are creative tools, and they serve creative purposes. The macrocreation force is the underlying force of existence, and it informs all aspects of our reality, including medications. If a particular creative purpose will be served by a particular medication proving ineffective, then this will be the case, regardless of any "scientific" indications.

Medical professionals will admit that they generally do not understand how medications work. Using various means at their disposal, it can be determined that certain medications will likely be effective in certain situations. When this relationship has proven sufficiently reliable, then the medication will be officially approved for certain uses. But the reason why one particular medication has the effect that it does in particular circumstances is usually not understood. Illness will always prove a major mystery to humanity so long as it is treated as a mechanistic process and not as a creative one. So long as illnesses have a role to play in our creative affairs, then cures for all illnesses will continue to elude us.

As stated elsewhere, the mechanistic processes that are at work in our world are at the service of the underlying creative one. This means that, many times, taking an aspirin or other common medication for a headache will generally alleviate the pain of that headache. In a general way, the medications that we use will have a fairly high degree of reliability. But this mechanistic reliability can be, and will be, overridden if there is sufficient creative cause. A medication that has worked well for a number of years may suddenly cease to do so, because the creative dynamic has changed and a new creative scenario has emerged.

Despite New Age dogma to the contrary, "negative" thoughts do not attract illness or injury to any individual. Energy, as has been explained elsewhere, exists in the form of a complex spectrum of characteristics, and it cannot be neatly divided into "negative" and "positive" halves. This does not mean that one's thinking patterns have absolutely no impact whatsoever on health issues, but there is no simplistic, one-dimensional relationship between one's thoughts and one's health. New Age types, and others, should not expect there to be simple solutions to their health challenges.

The healthiest people in our society are those people who understand the intrinsically creative nature of our lives and who thereby utilize creative means to deal with the issues and problems that inevitably arise. Rigid, one-track thinking, regardless of the dogma involved, is inherently unhealthful.

Situations of injury quite obviously bring up creative factors as well. For instance, tripping and falling can lead to a wide array of consequences. In this type of incident, you can emerge unscathed, or you can do major damage to one or more bones in your body. One tiny alteration in your trajectory as you fall can make a significant difference. It will be the result of macroself manipulation as to how you fall and whether damage is done. This manipulation is quite easy to accomplish at the macro level and routinely affects the course of our daily lives.

When we suffer from illness or injury, it is because a particular creative purpose is being served. There is no reason why knowledge of this dynamic should necessarily make our suffering any easier to bear, but at least we are free from feeling cursed by the unfathomable whims of a mythical deity. When we endure pain, it is not because "God" hates us, but because we live in a dramatic reality where there will always be difficult challenges to face. The more dramatic our lives are, the more meaningful and satisfying it will be at the macro level, which is where reality is designed. This microreality serves the needs of macroreality, and we all have creatively important parts to play.

Consciousness and Lack of Consciousness

In ultimate terms, consciousness can be equated with existence itself. Therefore, to be truly unconscious would be the same as being nonexistent. This means if you knock a person on the head, and they become "unconscious," they are still, in fact, conscious. The term "unconscious" does not describe anything that is real.

Regardless of what may happen to us at any moment from any cause, we cannot "lose" consciousness—all that can occur is a shift in consciousness. Awareness is a given fact in our lives. It cannot be diminished or taken away by any set of circumstances. We are aware because we exist, and we exist because we are aware. We will never experience a "fade to black."

The coma state is one area of consciousness that can be looked at. It is generally the case that when a coma patient awakens after a prolonged state of coma, the patient will usually exhibit a fair amount of lucidity and coherence. Even if years have passed, the awakened patient will show rational mental activity. If those years had passed in a condition of virtual nonexistence, it seems unlikely that such a strong degree of coherence would remain. During the time spent in the coma, the patient retained his awareness and lucidity.

Organized mental activity that continues in the coma state allows the patient to retain a continual condition of awareness, and this makes coherence upon waking possible. If the brain had "shut down" during that period, or had been engaged in endless hours of dreaming, then a dazed and confused condition would be the likely result. Even one night of rather chaotic dream imagery can leave the average person slightly disoriented in the morning. Several years of these mental meanderings would be quite difficult to deal with.

Generally speaking, awakened coma patients will recall very little of the mental activity that they experienced while in that state. Sometimes, they will report that they heard the voices of those who were gathered around them but were completely unable to respond. For the most part, however, they recall nothing at all.

Medical professionals take the fact that coma patients usually have very little memory of their time spent in a coma as an indication that the patient was "unconscious" and having little meaningful mental activity.

It should be understood, though, that the reality that the coma patient was experiencing was an alternative reality to this one—a competing reality that, for the patient's well-being, should not be recalled upon waking. Most of us can only handle one reality at a time, and so memories of others are suppressed upon waking.

This microreality that surrounds us during our waking reality is but one of an infinite number of realities that have the potential to be experienced. The vibrancy and intensity of this earthly realm derives from the fact that we are mostly ignorant of all others. We have roles to play in this world, and we cannot fulfill those roles adequately if we are devoting our attention to alternative realities. In the world of drama, an actor who "loses" himself in his portrayal of a particular character is usually going to give a more vivid presentation than an actor who is always glancing backstage and wondering if there will be champagne waiting for him in his dressing room after the show.

Even when we "lose consciousness" for just a few seconds due to fainting, a blow on the head, or some other cause, our awareness will instantly shift to another reality—one of the dream states that we have ready access to. To have even one moment of "blackness" would be the same as having one moment of nonexistence. Consciousness cannot be snuffed out any more than energy can be extinguished. There can only be a change in form or a change in awareness. Existence is absolute.

A patient in a coma is as fully "alive" as we are. This microreality has faded away from their consciousness and has become as vaguely remembered as our dreams are to us when we wake. They have a world to participate in just the same as we do. Their world fully involves them just the same as our world involves us. For the coma patient, the bulk of their awareness is focused in a reality that is an alternative to ours, but they are able to maintain a sliver of attention toward this world to maintain their link to a reality that has become a secondary one but still important emotionally.

The coma state is a creative choice that offers certain creative rewards and certain drawbacks. It is not the stuff of tragedy, no matter how sad it feels for the loved ones who must stand idly by. Whether the shift in awareness occurs as a result of a coma, senility, mental illness, or any other cause, one's awareness is never lost and it is never diminished, no matter what the appearance may be. There will be more discussion of these aspects of consciousness in the pages to come.

Dreams

The various conclusions that have been reached by sleep researchers over the years regarding sleep and dreams can be fairly thoroughly disregarded in the light of new concepts of consciousness. Sleep researchers engaged in simplistic, superficial observation of test subjects in laboratory settings did not take into account philosophical considerations relating to the nature and function of human consciousness. The macrocreation philosophy has new ideas to offer on the fascinating world that our dreams represent.

So many scientists have been so self-impressed with their presumed scientific "advances" that they have been quite unwilling to take into consideration any factors that would complicate the conclusions that they have reached. The subject of dreams simply cannot be dealt with in a meaningful way without considering the functioning of consciousness. For scientists to assume that so-called REM sleep equates to dreaming cycles and that non-REM sleep corresponds to a lack of dreaming is the most simple-minded conclusion imaginable.

The fact of the matter is that sleep equals dreaming—due to the nature of consciousness, it could not be any other way. As stated in the previous chapter, consciousness is intrinsic to existence. To be unconscious, in any manner, is to be nonexistent. Sleep researchers, of course, would not have taken any notice of a philosophical aspect to sleep such as this. They simply devoted their time to studying eye movements and imagined that they were accomplishing something meaningful.

When a test subject is awakened during periods of non-REM sleep, they will consistently fail to recall any dreaming that might have been taking place. Researchers therefore concluded that no dreams were taking place during these periods. However, non-REM sleep accounts for periods when our dreaming is performing a function where recall is simply not possible. The dream state represents a focus on a reality that is an alternative to our usual waking reality. In the case of REM sleep, this alternative reality is one that is sufficiently related to our waking reality that it is possible to integrate those experiences into our waking consciousness. In non-REM sleep, this is not the case.

Our dreams fall into two distinct categories—they either have a primary relationship to our microreality or they will have a primary relationship to macroreality. Dreams that take place in an alternative reality that has a strong correspondence to our waking reality have the potential to be remembered, whether they are or not. REM-type dreams take place in a world that, to a great extent, mirrors our own. This alternative reality is one that we can easily relate to.

The other type of dream, the non-REM type, involves our ongoing relationship to macroreality. These dreams cannot be recalled upon awakening because they correspond to a reality that is incompatible with our own—macroreality. These dreams routinely take us to the very same reality that we will encounter at the time of our "physical death;" for the entire time that we live out our lives in this microreality, we also exist and have awareness focused in macroreality. When we sleep, our "macrodreams" allow us to be removed from the great challenges and frustrations of this microreality and gain the insight and perspective that macroreality allows.

Our lives in this earthly realm can be so fraught with stresses, traumas, and tragedies that only routine sojourns in macroreality can provide us with the resources we need to cope. This world can take us to the very limits of our endurance, and we very much require the profound relief that time in macroreality allows. In the macro level of dreaming, we receive continuing guidance, reassurance, and respite from our microreality woes.

Imagine the relief that can be experienced in the macro state of dreaming by someone who is living life as a paralytic, in great pain and immobilized every waking moment, or someone who is facing starvation and the depths of poverty on a daily basis. Macro-level dreaming provides our best coping mechanism.

Obviously, having conscious awareness of the macro level of reality would have a huge impact on our lives, and this impact would, for the most part, be unhelpful. It is essential that we have the overall sense that this earthly realm is the primary reality of our existence. We need to cling to the "real" world as if it were the only one. It is acceptable that our awareness be something that we maintain a subconscious knowledge of—we know, yet we don't know, that macroreality is at the heart of existence. Thus, our dream experiences in macroreality cannot be recalled, even though it is vital that they occur.

This waking reality exists because all the life forms within it maintain the focus of their attention upon it. When we shift our attention, we shift our reality. Even a simple daydream can take us to another place, and the world that had surrounded us fades away. The waking reality is a collective reality—a group effort—while the dreaming reality is much more an individual reality. Submerged awareness of individualized realities allow this collective reality to have the forceful presence that it does in our lives.

Those dreams that we do have the potential to remember, those dreams relating to this microreality rather than to macroreality, can be seen as works of art that are crafted by our unique personal energy. These dreams are minidramas that relate to the dramatic nature of our waking lives. These are the dreams that can entertain, enlighten, and frighten us night after night. These dreams can be powerful enough to affect the course of our daily existence. Whether these dreams are consciously remembered or not, they do affect us. Biologically speaking, nothing that happens in our bodies or in our minds can be thought of as utterly pointless—we dream because we need to dream.

If we can remember our dreams clearly enough, they can tell us a great deal about the nature of reality. The dream reality is clearly a nonphysical reality, yet every single sensation that it is possible to experience in "physical" reality is possible to experience in our dreams. A recall of the dream reality that is crystal clear and "photographic" in its sharpness of detail is necessary to reveal these aspects of "physicality." Some people have had this highly detailed dream "snapshot" many times during their lives, but even one single pure memory of the dream environment is really all it takes to be illustrative.

In a dream, we can feel the golden warmth of the midday sun upon our shoulders, as well as the whispering touch of a cool ocean breeze. We can sense the precise temperature of the air or the water. When we walk down a sidewalk, we can experience the rough solidity of the concrete beneath our feet. All the "physical" nuances of the waking world are also present in the dream environment. There is nothing missing. For most people, the dream environment is a vague realm only because their recall of it is not sufficiently clear. But whether we are asleep or awake, there is only one thing that surrounds us—reality.

In the dream environment, we can also experience the spiciness of sausage and the sweetness of soda pop. Our taste sensations are real because our bodies are real as well. We can feel the achiness of tired muscles and the pangs of a hungry stomach. When we are in the dream environment, we inhabit a reality that is nonphysical yet has all the sensations that we expect to find in a "physical" environment. This is because there really is no such thing as a "physical" environment as that term is understood. The only reality that it is possible for us to experience is a nonphysical one. That is all there is.

Of course, the dream reality and the waking reality are not identical in all their attributes. The dream realm is fluid—scenes that we witness can mutate and change. This fluidity mirrors the nature of macroreality, which is the "central" reality of existence. The imposed rigidity of our waking reality is an aspect of presumed physicality rather than actual physicality, because there is no such thing as "physicality" in the way that our science texts define it.

When we do recall something of our dreams, we may be able to remember the general plot or simply the most dramatic moments of the dream experience. Very seldom does the dreamer recall the temperature of the air or other such specific details. It can be stated that dreams have a "dreamy" quality simply because our memories of them are so very poor.

Also, children are indoctrinated by adults to treat their dreams as trivial experiences that have no real meaning or significance. The statement, "It was only a dream," tells children that what they experienced during the night was not worthwhile. Because of this indoctrination, children learn to dismiss their dreams from their minds upon waking. Once this behavior becomes ingrained, it will be quite hard to un-learn it and to give our dreams the importance that they deserve.

Very few people in our modern, materialistic society make the effort to gain a sharply focused recall of their dreaming adventures. Generally, society will not offer any encouragement when it comes to the study and analysis of the dream environment. Such a useful device as "conscious" dreaming is given little credibility, and yet it is something that can help to transform an individual's ability to understand the world that we live in.

For some fortunate people, "conscious" dreaming comes easily and naturally. For others, it is a technique that can yield dramatic results if sufficient effort is applied. When you dream "consciously," you maintain an awareness that is comparable, or identical, to your usual waking awareness. Thus, when you awake, there will be a "bridge" of awareness that links your dreaming reality to your waking reality. When this bridge of awareness exists, there is the potential to remember the very smallest details of the dream environment.

With "conscious" dreaming, you can zero in on the tiny details of the world around you and observe what you wish to observe. It is quite possible to see and to remember such very small details as flecks of dust, grains in wood, patterns in a carpet design, each stone in a stone path, and even individual blades of grass in a lush lawn. This dream reality has all the elements that comprise our waking world—things that are wet feel wet, things that are solid feel solid, things that are hot feel hot. The snow that falls on us in the dream environment looks and feels exactly the same as the snow that falls on us when we are awake. The dream reality is reality.

The essential difference between the two realities is the fluid nature of the dream environment. But it should be kept in mind that the "physical" rigidity of the waking world is an artificial construct designed to meet certain creative purposes. The fluid domain of the dream environment is the most natural reality in existence—it is free-flowing and effortless. The artificial environment of our waking hours requires much creative effort to maintain.

Few people realize that the dream environment can provide the same degree of "grittiness" that our waking world offers. Walk down a city street in the dream reality, for example, and you can observe such "real world" details as litter-strewn lawns, patches of weeds, trash cans heaped with garbage, light poles, traffic signals, street signs, cars, trucks, and everything else that defines a scene as "real."

On that same street, we can observe every window in every building. We can even observe every pane of glass in every window in every building. With a brick building, we can, if we wish, observe every single brick. We can take note of tiles, cement blocks, ironwork, woodwork, colors, textures, and conditions. If we listen, we can hear the drone of traffic on the city streets and the overhead roar of a passing jet. With "conscious" dreaming, the dream environment can be seen for what it is: an alternative reality that lacks nothing in its full expression of what we consider to be "real." Without sufficient recall of our dreams, all these details are lost, and a dream becomes merely a dream.

The dreams that have the potential to be recalled use the imagery and sensations of waking reality and combine this with the natural fluidity of macroreality. The imposed rigidity of our waking reality is quite a burden for us to bear, and our dreams give us much-needed relief from it, recharging us and reminding us of what the "real world" is all about. Our dreams provide us with a means of escape from the imprisonment of this microreality.

Regardless of what we recall of our dream experiences upon awakening, every moment of our dreams is retained in a particular place in our memory storage. These moments may or may not emerge again into our waking consciousness, but nothing is ever lost. For instance, if you have a dream that takes place in New York City, many of your previous dream experiences that took place in New York City may surface to your dreaming consciousness. Thus, you will have a full "history" of your dream self in New York to draw upon, and that will add to the richness of the dream experience. The key to this memory retrieval is association—in this case, the category is New York City.

Because our waking recall of the dream environment is usually so poor, we do not realize that our dream self has its own wealth of memories and associations to draw upon. The dream self knows that it is a real self, and it is, in every way.

In conscious dreaming, the dream self can maintain an awareness of its waking self, yet not discount its own sense of legitimacy. In our waking world, we are basically limited to portraying one character throughout the course of our lives. However, our dream self knows no such limitations and can cast itself in whatever "part" is the most creatively satisfying at any moment. This, too, mirrors the world we will know in macroreality.

Although it is possible for us to achieve an absolutely exact duplication of our waking reality in the dream reality, we will generally opt to create large or small variations of familiar locales. This is because it is naturally much more creatively satisfying to create something new than it is to duplicate something old. Of all the painters in the world, only a very few find the technique of "photographic" realism very fulfilling. Thus, when we dream of our own home, for example, we will create various alternative views of it simply to satisfy our own creative whimsy. When we dream of a town or city that we are familiar with in our waking lives, then the chance for vibrant creative expression is usually impossible to resist.

Despite what some people may believe, our dreams do not serve primarily as a message delivery system for our own subconscious minds. The macroself and our human, waking self are in constant communication throughout the day, and this is the case whether we are asleep or awake. This contact is an automatic process, like our breathing, and the dream state is not necessary in any way for this connection. The complete self is an aspect of a continuum of personalized energy, so one self is never "cut off" from another self. Whatever messages that are contained in our dreams will take a certain form, and this form is intrinsic to the dream state, but we are always in a position to receive guidance and information from the macroself.

Sometimes, the messages that are contained in our dreams will be readily comprehended by the waking self in an overt manner that lets us know that the message has been received. Generally, however, the messages that are contained in our dreams are received and comprehended without us having any conscious awareness of it. We do not need to make a special effort to determine what our dreams are telling us unless we have the sense of being uneasy and bothered by it. In some cases, our dreams may disturb us so as to affect our waking lives in a particular way. Generally, however, we can assume that some part of our self is always in a position to receive and interpret these messages from the macroself. Our dreams are an aspect of our biology, and biology operates in a very logical and functional manner.

When we are living our waking lives, we are confined to the unyielding imprisonment of "physicality." When we are dreaming, we are free. Our nightly flight of freedom is a necessary escape from the straightjacket of our waking hours. Our dreams remind us, whether we realize it or not, of the natural fluidity of existence. To dismiss our dream experiences as trivial is to dismiss life itself as trivial.

Recreational Drug Use

The use of recreational drugs has been highly stigmatized in most societies around the world. The primary reason for this has to do with humanity's innate desire to have common agreement regarding the boundaries of reality and unreality. Users of recreational drugs may often develop an individual perception of reality that can differ from the agreed-upon parameters that society at large has chosen to enforce. Thus, society develops a cadre of "reality police" whose job it is to come down hard on those members of society who seek to achieve a unique perspective on reality through the use of mind-altering substances.

In fact, individuals who do form independent perceptions regarding the nature of reality do pose a certain small threat to the established order, but this is a threat that should be tolerated for the good of all. Human society is indeed strengthened by the agreed-upon framework of reality that it develops for itself. "Reality renegades," such as drug users, are indeed able to shoot holes in this framework. However, a truly enlightened society will have sufficient awareness of the truly complex nature of reality that it will be able to cope with any challenges that individualistic perceptions may pose.

At present, our framework of reality has been in the hands of fundamentalist types, both religious and scientific. People with fundamentalist tendencies tend to exhibit high levels of personal anxiety and thus require a simplistic formula to explain the world that they are a part of. A one-dimensional, black-and-white framework suits them best. When their fundamentalist framework is challenged by those who see things differently, the reaction amounts to a complete rejection of any alternative view, often with a touch of mild hysteria. It is human nature to prefer simple lines of reasoning and simple explanations for what is going on in this world. Recreational drug users are vilified because they have perceptions and perspectives that are guaranteed not to fit in with a fundamentalist framework.

Perhaps due to the fact that society's framework of reality is so rigid, many people do have a low-key yearning for a long-ago world of myths and fairytales—where a "magical reality" was the order of the day. This was a time when fire-breathing dragons walked the earth, wizards cast their spells, and warlocks chanted their incantations, and when giants, fairies, trolls, and unicorns all enjoyed their role in the grand scheme of things. This is an epoch that understandably appeals to the child in all of us.

Perhaps those mythical days of yore were not mythical at all. Perhaps some sort of magical reality was in operation in the distant past—a reality that was distinctly different from the stringently rigid framework of reality that we are accustomed to today. One possibility to explain the strong role that recreational drugs play in the world today is that many people have an instinctive yearning for a time when this world could provide pleasures that are unavailable today.

We live in a world that has become increasingly defined by fundamentalists, both of the religious and scientific persuasion. These simple-minded people prefer simple-minded explanations and structures to frame their lives. The concept that someone who cast a spell could have an actual impact on the affairs of this world causes them far too much anxiety to live with. The population at large has allowed these strident, self-glorifying loudmouths to take up the prominent position that they now hold in human society.

The enduring popularity of science fiction and fantasy fiction demonstrates a profound human desire for a world that is more purposely imaginative than the one that we endure in our current era. The occult arts and sciences, which are undeniably irrational in their forms and formulas, have nonetheless endured throughout human history due to their appeal to our imaginative senses. Occult practices can be debunked but can never be destroyed because of the necessary creative role that they perform in our world. Fundamentalists simply do not comprehend this.

Recreational drugs help the individual to escape the tyranny of the fundamentalist reality framework. Of course, there are many different kinds of recreational drugs, and they provide a wide variety of effects on the user and on society as a whole. These drugs, in general and in varying degrees, provide a fluidity to personal reality that is a welcome relief from the framework of concreteness provided by the fundamentalists, particularly the scientific ones. Most human societies in the world do not approve of this escape hatch, because it threatens the imposed order, but no society has the power, or the authority, to prevent the use of recreational drugs by a significant percentage of the population.

Society, for the most part, will tolerate mild drugs that it perceives as no great threat to the established order. Despite the widespread impact that these "mild" drugs can have, society has a general recognition that alcohol, nicotine, and caffeine do serve to help many people to face the harsh challenges of this world and "make it through the day." These particular drugs are not perceived as "subversive" and are therefore permitted. Almost all other recreational drugs are considered subversive by most human societies and are, therefore, forbidden.

One "mild" drug that is not well tolerated by most societies is marijuana. The central reason for this is the perceived notion that marijuana users have the tendency to sit back, relax, and "drop out" of society's mainstream. In our fast-track capitalistic marathon, "dropping out" is considered quite subversive indeed. Society cannot endorse a drug that seems to have such a negative impact on worker productivity levels. Alcohol, nicotine, and caffeine have all found a home in the modern workplace, but marijuana never has. Therefore, marijuana is "wrong," while other mild drugs are "right."

A common drug that plays the opposite role of marijuana is cocaine. Because cocaine is understood to be a go-go stimulant, it is much more in line with a capitalistic society's goals and attitudes. Cocaine is seen to be an aid to energetic individuals who desire to be even more energetic. To a certain extent, this strong drug receives society's covert blessing. An individual who wants to both work harder as well as play harder is not perceived as posing a threat to society. Someone in the public eye who has been found to be a cocaine user seldom pays a terribly high price, especially if they have had achievements that society approves of. Only low-class "pushers" are likely to receive public condemnation.

Heroin is probably the most demonized of all recreational drugs. There is no question that the heroin user has a very strong likelihood of falling out of society's mainstream. The infamy of heroin derives not from its addictive qualities, which are undeniably powerful, but from its anti-social aspects. The heroin user is intrinsically a rebel, and most societies tend to put down its rebels, no matter what they are rebelling against.

While there is no justification for glorifying those who choose to use powerful recreational drugs such as heroin, there is no possible justification for demonizing these people either. A user of heroin is seeking a personal reality that will have more meaning for them than the general concrete reality that society promotes. This personal reality may, in some people, also lead to personal destruction. Thus, society pretends to have the right to "protect people from themselves," but the real reason such strong drugs receive society's condemnation is self-protection. Society will seek to punish those individualistic seekers who wish to break away from the capitalist society's manic pace and become "non-achievers," which is a terrible sin.

Society also has no tolerance for another class of recreational drugs—the hallucinogens. If it were not for the fact that very, very few people consume this type of drug, it would be labeled Public Enemy Number One by the reality police who serve society's interests. Hallucinogenic drugs promise to deliver the user to a personal reality that is innately free of society's strictures, and nothing could be more subversive than this. Even when a drug such as peyote is used in religious ceremonies that date back for numerous generations, society cannot tolerate it. It takes a brave individual to withstand the challenges that a hallucinogenic drug has to offer, and society should honor these individuals rather than persecute them. We live in a society where such inane accomplishments as climbing a mountain are heralded and rewarded, yet we opt to demonize those souls who take on a far more significant challenge—seeking out alternative visions of existence itself.

There are many other drugs in use by various members of society that may have less redeeming value than do the hallucinogenic types, but these drugs should be tolerated by society as well because they offer the individual the chance to craft a personal reality, and this should be considered a basic human right. The fact that this choice of drug use may ultimately lead to the user's destruction is a sad thing, yet it is a human being's intrinsic right to follow his own path. Society does not need to offer its approval, only its reluctant tolerance.

We currently dwell in the Dark Age of concreteness, but this medieval era will eventually give away to a magical renaissance of fluidity and individualistic reality. Society will no longer live with the unreasonable fear that its bulwark of concrete constrictions will come tumbling down, throwing the world into chaos and confusion, simply because some members of society choose to alter their individual perceptions. Society need not ever embrace the use of recreational drugs, but it does need to understand the underlying motivations that drug users have. The definition of reality does not belong to society as a whole, it belongs to the individual.

Mentall Illness

Because many of the people who are considered mentally ill tend to be nonproductive members of society, this condition is very highly stigmatized. The mentally ill are on a path that diverges sharply from society's mainstream, and thus they are marginalized and denigrated. However, this need not be the case. A more enlightened society would learn to appreciate the contributions that the severely ill can make. The alternate perceptions of reality that can be an aspect of mental illness would be seen as potentially valuable.

Society, for the most part, has not begrudged the vast expenditures that have been devoted to the space program in various countries. These expenditures have helped us all to achieve a broader perspective on the nature of this earth reality. A schizophrenic, for example, is already in a position to explore regions that exist beyond the usual boundaries of our reality, and he did not have to spend a single cent to get there. Schizophrenics and others with severe mental illness are as much explorers of "outer space" as our astronauts are. These explorations into outer psychic realms need to be recognized as potentially valuable, and those people who are uniquely equipped to make the journey should be valued for what they have to offer.

"Normal" members of society—the harried mainstream—cannot go where the severely mentally ill can go. This means that the severely mentally ill and users of hallucinogenic drugs are a resource that others can draw upon. The fact that some marginalized people are not marching in mental lockstep with the majority is something that should be fully appreciated for the alternative insights and perspectives that can be offered. It should also be noted that if the mentally ill understand that they may be able to make some unique contributions to society because of their illness, then their sense of self-worth would be greatly enhanced and this should have significant therapeutic value.

A society that scorns variances of perception is an inherently weak society. A society that promotes a compulsive rigidity of philosophy is a society that is doomed to fracture. The severely mentally ill can be helpful in fostering a more multi-faceted framework of reality that can have positive implications for society as a whole. No one else can go where they can go.

The macrocreation philosophy contends that all perceptions must be considered "real" because all perceptions, no matter what their nature, represent the expressiveness of energy, and the expressiveness of energy is the essence of existence. This means that the "visions" and "hallucinations" that can come with severe mental illness should not be dismissed out of hand by society. Those individuals who experience unusual perceptions should be encouraged to provide accounts of their "journeys," and these accounts should be retained and analyzed. While there is no guarantee that these hallucinations and visions will be of any great worth to society, there certainly is no guarantee that they will not be, either. The information that is provided, whatever it may be, should be taken seriously for whatever value that it may eventually prove to have.

If all members of society share a fairly narrow range of mental perceptions, then this would have to be considered a limiting factor. In terms of normal biology, the gene pool needs to be sufficiently varied in order to offer the requisite environment for producing healthy offspring. In terms of human philosophy, the "perception pool" also needs to provide sufficient variations. Without those variations of perception, our philosophical point of view will become constricted and "inbred." If every member of society perceives reality in a rigid, prescribed manner, then it can be compared to every single person marrying his or her first cousin.

Because existence can be defined as the utterly intricate interchange of all the characteristics of energy, there can be no sharply defined demarcation between what is labeled "real" and what is labeled "not real." The divisions between "real" and "not real" that society has established certainly have practical value, but have little underlying value in philosophy.

The nature of reality also pertains to what we label "subjective" and "objective." Perceptions that are entirely individualistic rather than communal are given no credibility by society. "Rogue" perceptions, whether as a result of mental illness or drug use, are denigrated and ridiculed. In support of group-centric reality, heretic perceptions have been scorned so as to reduce the possibility that they will have any impact upon the agreed-upon boundaries of reality. Thus, the mentally ill are marginalized to the very edge of society, and nothing they say is given any credence whatsoever. Society, as a whole, is simply too afraid of what the mentally ill may have to report from the journeys that they have taken.

Whether a perception is individualistic in nature or communal has no bearing on the validity of that perception. All existence is the interactions of patterns of energy, and there simply is no reasonable way to categorize those energy interactions as being "real" or "not real." The mentally ill can perceive things that others cannot. This is because they have the personal mechanism to do what others cannot do.

Scientific fundamentalists have kept the spirit of seventeenth century Salem alive in the twenty-first century. They have fulfilled their role of "reality police" with the lust typically found in the most ardent missionaries. Scientific fundamentalists will burn heretics at the stake of public ridicule. They have reserved the right solely upon themselves to declare the boundaries of our mutual reality. As far as they are concerned, those reality renegades who choose to sail in uncharted waters in search of new lands will most likely fall off the edge of the world.

Another aspect of mental illness that can be examined is the occurrence of senility. As a result of common senility, as well as Alzheimer's and other forms of dementia, many of these sufferers give the impression of being "not all there." In terms of the nature of consciousness, this may be literally true.

For creative reasons, the personalized energy that forms the matrix of human personality can be gradually withdrawn from this earth reality. The death of an individual in our microreality can occur instantaneously or it can occur in stages. If the dying process is quite gradual, then this represents a slow immersion into macroreality and a slow extrication from this microreality. For many people, this becomes the desired "way to go."

The human personality never disintegrates, despite any appearance to the contrary. If the personality of someone we know appears to be significantly diminished, it is because a degree of that person's individualized energy has already departed this microreality and is focused elsewhere. Thus, no actual diminishment of the total personality has taken place—instead, there has been a re-allotment. The degree of personality that remains behind in our reality can be so minimal that it may leave nothing more than a "shell" for us to interact with. Of course, a shell personality will not be a very satisfactory one for the loved ones and caretakers to deal with.

In many cases, the bulk of the energy that comprises a human personality will depart this world a long time before the physical body does. This is because the timing of death is a creatively critical event and has repercussions on a great number of people. It is therefore possible for a person to "die" years before the physical body does, utilizing the transference of individualized energy from one reality to another. No one is ever condemned to existence in a state that can be labeled "vegetative," regardless of appearances. If the fullness of an individual's personality is not present in our reality, then it is present in macroreality.

While senility, dementia, and coma are heartbreaking for the sufferer's loved ones, it is a much gentler process for the impaired person himself. The senile person is making the transition from our reality to the next one in a very leisurely manner. When the time for complete physical death finally occurs, the transition will already have been virtually complete. The timing of death serves the needs of the deceased, but it also serves the needs of all those who know him or her.

For example, if a significant inheritance is involved with the estate of the deceased, that money may have a rather profound impact on the life of the beneficiary or beneficiaries. The new funds may very well change the course of the person or persons who are the recipients, and not necessarily in a simplistic, "lucky" way. Acquiring new wealth can be quite disruptive and destabilizing, especially if the beneficiary is young and/or immature. It may make a significant difference if those funds come to someone at the age of thirty rather than at twenty-five.

While an elderly individual may be perfectly ready to "go" at age eighty—quite content to have his earthly existence concluded—the impact of his death on his loved ones at that particular time might be problematic. Thus, an extended period of senility, dementia, or coma may allow the elderly person to depart this world when he is ready to, leaving his body and a mere shell of personalized energy behind.

When the timing of death is best for all concerned, the final act can take place.

Mental illness, in all its forms, is an important creative device that has much dramatic potential in our world. Any mental illness poses a major challenge to the sufferer, and this microreality is designed to be an arena of challenges of all sorts. Those who endure mental illness should be given full credit for the difficulties they are facing and for the alternative perspectives they are achieving. Rather than dwelling in hopelessness at the margins of society's field of focus, the mentally ill should be front and center—honored for their uniqueness and courage.

Macrocreation Glossary

Abstract Core
This is the primordial, utterly nonhuman realm of wordless abstraction that lies at the very center of existence. This region is as far removed from humanity as it is possible to be.

Associated Self
This is the personalized energy that exists along the continuum of energy that has an intimate "familial" relationship to our own energy matrix. These associated selves have been mistaken for "reincarnational" selves in some religions and New Age philosophies.

Common Natural
These events, occurrences, and conditions occur in a mechanistic manner in this earth reality. The common natural comprises all that we consider "normal" as regards all the scientific "laws" that we rely upon. The common natural is to be contrasted with the "uncommon natural."

Concrete Extreme
This is the rigid, unyielding "physical" nature of this earth reality. It is the opposite of "fluid" reality.

Energy Landscape
This planet's "physical" landscape can be understood as an energy landscape in terms of the qualities of energy that comprise existence. Every spot on the energy landscape has a unique energy signature.

Energy Matrix
This is the focus of personalized energy that comprise a being, whether it be a human being, or otherwise. Another term for this is the "energy self."

Energy Signature

This is the complete nature of a being or an aspect of life that is depicted in a concise expression of its energy aspects. Conceivably, an energy signature could be expressed in a recognizable manner, such as digitally.

Energy Spectrum

This represents the totality of all possible attributes of energy. It can be understood as possessing qualities of "coloration" that relate to aspects of life in our world, particularly qualities that human beings label as "good" and "evil."

Macrobeings

A macrobeing is an energy self whose primary focus of existence lies in macroreality. An example of a macrobeing is an angel or a demon, both of which can be defined in nonreligious terms and relate to the creative nature of energy.

Macrocommunication

This is the fundamental means of communication between all forms of life. Communication that utilizes human language rides upon the surface level of macrocommunication.

Macroconscious

The macroconscious is the awareness that accompanies the energy expressions of existence itself. Although the concept of the macroconscious equates to the mythological portraits of "God" that human society has produced, the actual nature of the macroconscious should not be confused with those portraits.

Macrocreation Force

This is the fundamental impetus that motivates energy to be active and, in conjunction with the macroconscious, to be organized. It is the only force at work in existence.

Macrocreation Philosophy

This is the belief system that applies the trinity of principles of the macrocreation theory to all aspects of life. The macrocreation philosophy is created on an individual basis by anyone who adopts the principles of the macrocreation theory. Thus, a macrocreation philosophy will involve some subjective judgments and intellectual conclusions that may have little appeal to those who have also adopted the trinity of principles of the macrocreation theory but who have come up with their own version of a macrocreation philosophy.

Macrocreationist, Macroist

These terms describe a person who has adopted the principles of the macrocreation theory. This person will then create a macrocreation philosophy that appeals to their own particular intellectual inclinations.

Macrocreation Theory

This is the theory of existence that is solely composed of the trinity of principles that pertain to that theory. The corollary topics regarding the nature of time and the question of existence versus nonexistence are not formally a part of the macrocreation theory.

Macrodream

This is a dream that has a primary relationship to our existence in macroreality. This is a dream that, under ordinary circumstances, cannot be remembered upon awakening.

Macrolanguage

This is the wordless, symbolic, and complex language that is utilized in macroreality and may also find expression in our dreams. It is a language that is generally too sophisticated to be fully comprehended by the conscious human mind.

Macroquotient

This is the measure of the ability, or lack of ability, of a human being to achieve a perspective that is much broader and more objective than the usual human perspective.

Macroreality

This term applies to the totality of existence, but it can also be used to describe the world that exists outside our earth microreality.

Macroself

This is the totality of the personalized energy that comprises the self, but especially the energy focused in macroreality.

Macrotime

This is the nonlinear nature of time that is in operation in macroreality. In our earth reality, we utilize a linear system of time that is the result of program design.

Microdream

This type of dream relates to our earthly experiences and has the potential to be remembered, whether it is or not.

Microreality

The earth reality exists as a subreality of the greater macroreality and is generally the only reality that we have awareness of.

Reality Police

This term refers to those members of society who have taken it upon themselves to support a communal, standardized framework of reality and who will enforce this framework by whatever means are at their disposal.

Reality Renegades
These people prefer a more personal and individualized understanding of reality that may be in distinct variance with the standard, communal view.

Scientific Fundamentalist
This is a person who combines scientific ideas with a personal fundamentalist mindset. The psychological rigidity of the scientific fundamentalist is often mistaken for a rigorous application of scientific principles, but this is simply the appearance.

Trinity of Principles
This is the set of principles that constitute the macrocreation theory.

Uncommon Natural
This term can be used as a replacement for "supernatural." Incidents of the uncommon natural may be quite rare in the frequency of their appearance, but are nonetheless perfectly "natural."

Section Four—Personal Philosophical Development

Opening Remarks

This section will provide some autobiographical details relating to the development of the macrocreation philosophy. This account will show the very gradual process that was involved in the formulation of my ideas and my ultimate philosophical conclusions.

My formal education in the study of philosophy was quite meager. There was one course covering well-known Greek philosophers of the classical era, and that was it. Because I found myself rather unimpressed with these famous thinkers, I never pursued further formal study. Not only did I choose to ignore the Greeks, I also opted not to read the great German, French, and Chinese philosophers as well. My personal philosophical development has been truly **personal** in that my primary interest has been pursuing my own line of inquiry, rather than becoming expert in other people's ideas.

Therefore, for many years, my philosophical development was the result of querying myself and very patiently waiting for any insights that might eventually come. This process allowed for a very slow aggregation of insights and ideas—like pieces of a large-scale mosaic, a clear picture gradually took shape. There was never a single instance of "blinding" revelation. At no point in this long, slow process did I believe that I was on a philosophical mission for the benefit of humanity. My quest for philosophical understanding was entirely self-centered. As I developed these ideas over the course of twenty years or so, I never thought that I would make any effort to spread my personal philosophy to the world.

Ultimately, I did decide that writing everything down in book form might be a worthwhile endeavor. I realized that there was a possibility that these ideas might be useful to someone besides myself—people who lacked a religious, New Age, or scientific belief system that fully answered the questions that they had about this world.

Every human being who has ever lived can be considered a philosopher—living and thinking are the only credentials that matter. The macrocreation theory and the macrocreation philosophy are the result of my living and my thinking.

Memoir and Philosophy

For most people, the primary philosophical influence in their lives comes from their parents. Far too often, a child's views will turn out to be nearly identical to that of the parents, which indicates that there has been no personal philosophical development taking place at all. In my own case, my parents were certainly influential, but I charted my own course in life from a very early age. Since my parents were very different people, they were very different in the way they influenced me. Growing up, I learned to think quite independently because they were pulling me in different directions.

My mother was a fairly unemotional person, a rationalist who embraced the modern scientific perspective. Although she had been raised a Baptist, the Christian religion did not enter her heart the way it does for many people. At one point she did force my brother and me to attend a local Congregational church, but this was mainly because she thought that it would be a good thing socially. She was not hostile to religion, but she always maintained a wary skepticism, and this certainly did have some impact on my own burgeoning beliefs.

In contrast, my father was a great believer in the "spirit world," and in all things supernatural. He possessed very little natural skepticism and had very little objectivity about his own life. My parents were both born in the Midwest but met and married while living in New York City. There must have been tension and conflict in their marriage from the very beginning due to their basic incompatibility on many levels. They separated when I was three years old and divorced several years later.

Shortly before my birth, my parents had relocated from New York to a small, pretty town in Michigan. My mother thrived in the new environment, becoming a successful local journalist and editor. She became an active clubwoman and served as an officer in several groups. Her charming and intense personality won her many dedicated friends and some prominence in the community. When she died suddenly in 1985, there was a large attendance at the services.

My father, on his own, did not do nearly as well. He suffered numerous "nervous collapses" and endured chronic depression, accompanied by debilitating anxiety. As a young artist, he had experienced success in both Los Angeles and New York. But in Michigan, his career as a commercial artist faltered. He realized that he had failed to fulfill his early promise. When he was eighty-four, he attempted suicide, and he was very nearly successful. He was a resident in various mental health facilities after that.

From my mother, I inherited my rational skepticism and "no-nonsense" attitude toward life. Despite her vivacity, she was a fundamentally serious person who plowed through each day regardless of how she felt. From my father, I acquired a high degree of sensitivity toward the world around me. He helped instill in me a love of nature, as well as an appreciation for the intriguing mysteries of the supernatural. Thus, I grew up as a rationalist with a poetic sensibility.

As previously mentioned, I did attend church for several years at my mother's insistence. The Congregational church that I went to was a fairly liberal one, and the Sunday school lessons were quite mild in their tone. Like my mother, I was not at all hostile to religion—I very much appreciated the beauty of the sanctuary and church building, and I liked the warm, loving atmosphere that the members of the church provided. But at that time, nothing that I experienced inside the church made much of an emotional or intellectual impression on me. I neither believed Christian theology nor disbelieved it. The Bible was merely a book of stories. When my mother ceased to require my regular attendance, I ceased to go.

Growing up, I read many books on supernatural topics. My favorite topic was death—ghosts and the afterlife. True accounts of hauntings gave me the most pleasure. The more horrific the haunting, the more I enjoyed the tale. Psychic predictions were also of interest to me, and I wanted to believe that it was possible to predict the future. I spent time exploring the occult sciences—astrology, I Ching, palmistry, numerology, and all the others. They were all fairly intriguing but never thoroughly convincing.

Because ghosts interested me so much, I am very glad that I had an encounter with one. My grandmother's house in Iowa was not a spooky Victorian mansion like in the movies, but it was a haunted place. When my mother was a little girl, she saw what she took to be a benign presence there, as it seemed to be guiding her away from a possibly dangerous space during a remodeling project. However, the presence that I experienced there was not so tranquil.

It was my custom to spend two weeks or so every summer at my grandmother's house. Iowa felt like a different world from Michigan, and I usually enjoyed my visits. However, routinely, I would be given a guest bedroom that always felt just a little bit eerie to me. Very often, I had the sense of being watched by unknown eyes. It was unnerving, but I grew rather used to it. Acting as my own parent, I would tell myself that it was "just my imagination." But the feeling persisted year after year.

One summer when I was in junior high, while occupying that spooky bedroom, I came down with severe but generalized pains in my stomach or adjacent organs. My mother took me to the hospital where I was admitted and remained for several days. Tests were run, but nothing showed up. The pains gradually lessened. The only diagnosis the doctor could come up with was "nerves." But I had never suffered an attack like that before, and I never did again. So, it may have been a ghost that was making me so "nervous."

When I was seventeen, my mother and I journeyed to Iowa for Thanksgiving. We had a very nice feast prepared by a relative in a nearby community then returned to Ottumwa where my grandmother resided. It had been an entirely pleasant day, and I had no idea that a "supernatural" attack was waiting in the wings.

Lying in bed in the "haunted" bedroom, asleep, I was jostled awake by the violent shaking of my queen-sized bed. The shaking was so rough that it was all I could do to hang on. Although I was terrified, I also reacted with a certain amount of anger. This innate anger, which has colored my life, helped fuel my response to the ghostly attack.

I was able to sense the location where the energy of the entity was focused. Gathering all my own personal energy, I struck at the invisible ghost with my fist. With this action, the shaking immediately ceased. I had fought a ghost and came out a winner.

For the rest of the night, I lay in bed victorious, but deeply shaken. When I came downstairs for breakfast, I must have still been rather pale because my mother looked at me and remarked with a light laugh, "You look like you've seen a ghost!" I smiled weakly but did not respond since my grandmother was there and I did not want to alarm her. My mother and I had breakfast, then drove back to Michigan. As it turned out, I never spent another night in that house.

Because ghosts do not fit in with the accepted dogma of scientific fundamentalism, their existence is denied. There is nothing innately irrational about ghostly manifestations. Scientific fundamentalists are people who have adopted a rigid belief system that is in accordance with their rigid personal psychological state. Because scientific fundamentalists proclaim their rationality in a loud and authoritative manner, they are taken far too seriously by many members of society, and this has been harmful to a truly rational philosophical inquiry on the subject. Ghosts have a bad "reputation" among self-professed rationalists, but it is an undeserved reputation.

My encounter with the bed-shaking ghost showed me that a "supernatural" being can be dealt with in an aggressive and forthright manner, as opposed to simply reacting with unreasoning fear. My instinct was to fight back, not to cower, and this attitude served me well. I knew, somehow, that I possessed a personal power that gave me options in this confrontation. There may have been some risk involved, but it was a risk worth taking.

However, in my next engagement with an unworldly entity, I chose a different course of action, and that may have been the right choice for that particular circumstance. The event took place on a darkening summer evening in my hometown in Michigan. It was a beautiful evening for a walk, and I had been strolling around for an hour or so when I came upon a cat in a bushy, shadowy stretch of lawn. The cat was sitting on the grass with its back toward me. Being a great cat lover, I decided that I would go over to it and say, "Hello."

The cat turned around and faced me. Its broad, evil grin caused me to jump up in the air in fright. My feeling was that this incident was no accident—this cat knew who I was and had come to this spot to give me a scare. But I chose not to think about the implications of this too much. With very little hesitation, I backed away then ran as fast as I could.

A scientific fundamentalist has no options when dealing with an account such as this. Because incidents of the uncommon natural do not fit in with their rigid, unyielding dogma, they can only conclude that this type of event simply did not happen. It was my imagination or a trick of light or I was inebriated or some other "plausible" explanation had to be the case. A fundamentalist of any type is boxed in intellectually and psychologically. They have no room to maneuver. They can only say one thing and believe one thing. Their behavior cannot be considered rational.

The incident in Iowa was the only time I have seen a ghost, thus far, and the incident in Michigan was the only that I have met up with a "demon cat." The rarity of these occurrences has no relationship with the validity of such occurrences. Events of the uncommon natural may be quite rare, inconsistent, and impossible to satisfactorily reproduce in a laboratory setting. Scientific fundamentalists demand consistency in the things that receive their stamp of approval because consistency corresponds to their simplistic belief structures. They have great difficulty in dealing with phenomena of a highly creative and complex nature. This is simply quite unfortunate for them, because this reality that we inhabit is intrinsically creative and complex. Those things that can be verified in a laboratory only have a limited usefulness.

My adolescence was a very uneasy one, mainly because my relationship with my mother had deteriorated to the point where we could barely communicate with each other. I was hostile, sullen, defensive, and disrespectful—year after year. This continued even past high school and into my early twenties. We shared an apartment, but we shared very little else. We were both quite miserable.

But this interpersonal dynamic was to change in a mere twinkling. One particular night, I had shut myself up in my bedroom as usual and was not doing anything that was the least bit unusual. Suddenly, I found myself soaring skyward, free of my body, attaining a great height above the town I lived in. With this dramatic new perspective, I acquired a viewpoint that was transcendental and transformational.

As I hovered in the tranquil evening sky, all my problems seemed much smaller and not nearly so overwhelming. I felt a gentle calmness wash over me and cleanse me of my petty resentments toward my mother. This was, in certain respects, something that could be considered a "religious" experience, yet there were no trappings of religion to be found. Neither God nor his angels made an appearance. There was no heavenly choir. This experience was everything that it needed to be to change my life, but it did not need to be religious, and it was not.

When I returned to my usual awareness, back in my own room, there was one compelling thought in my head—that I strongly desired to treat my mother differently from that moment on. I no longer wished to say things or do things that would make her unhappy. My grievances against her had not gone magically away, but I was resolved to put them aside and without expecting anything back from her in return.

From that moment on, I did, in fact, treat her differently, and our relationship subsequently improved considerably. This transcendent event occurred, I believe, at the macro level, where the path that I was on was seen to be harmful to both myself and my mother and a major, abrupt adjustment was needed. This significant course correction was accomplished in an expeditious manner through a transformational experience. While it is unusual for the macroself to operate in this manner, it is a tool that is at its disposal when the situation calls for it.

Although, at this time in my life, I was still very much open to the possibility of religion in my life, religious trappings were not to be found in my life-altering encounter. Since I had no deep-seated psychological need for mythical and religious ornamentation in my life, it did not occur. There was no need for the appearance of "God," and so God was not there.

In the 1970s, presumably for reasons of my particular body chemistry, I grew to be quite dependent on caffeine, generally a rather mild drug. Although I did not like coffee and never drank it again after my first taste of it, I was very fond of caffeinated soda pop. Typically, I would have some cola or other soda pop from the time I arose in the morning and continuing frequently throughout the day. After awhile, insomnia developed, but I did not cut back. The stimulative value of caffeine was so desirable that I was willing to put up with some unpleasant aspects of my excessive consumption.

However, by late in the decade, I was reaching a point where caffeine was becoming a big problem for me. I do not know if caffeine-related hallucinations are part of the literature on this particular drug, but it did happen to me. For a year or more, whenever I closed my eyes, I saw an eye staring back at me out of the darkness. It did not seem especially menacing, and I did not associate it at first with the caffeine consumption, so I kept on drinking. The eerie eye certainly did bother me a bit, but I managed to get somewhat used to it.

Then, in April 1979, I experienced a night of truly alarming hallucinations that had a major impact on my life. The single eye that I had been seeing became two eyes, and those eyes belonged to an iridescent green lizard, which I beheld whether my eyes were open or closed. Then, it grew far worse. My entire field of vision became filled with these iridescent lizards, all staring back at me, which I saw with my eyes open or closed. I lay on my bed in my dark bedroom and felt that I was surely losing my senses—these lizards had taken over my world. It was a demonstration of creative power that a shaman would have full appreciation for, but I was simply terrified.

The light of dawn chased the hallucinations away. I knew, somehow, that the years of overdosing on caffeine had led me to that nightmarish encounter. Of course, I resolved to give up caffeine from that day forward, and I did. But there was more to it than just that. The main reason I had abused caffeine was because its stimulative value helped me to deal with my chronic mild depression that had been a part of my life for a long time. I knew that I needed to adopt a more positive attitude toward life if I were to avoid the lure of stimulants in the future. This, too, I accomplished.

As I look back on this experience now, I can see that the appearance of the lizards had a meaningful symbolic value. It is not easy to interpret this symbolism, but there is a part of my awareness that understands it perfectly and took the lesson to heart. Instead of reacting with fear, as I did, I should have reacted with respectful wonder.

In May 1979, one month after my night of hallucinations, my life changed dramatically. Encouraged by a new sense of optimism, I departed my small Michigan hometown and took up residence in the very big city of Chicago. It was a bold step for me because I had very little money to my name and my entire life in Chicago hinged on getting a decent job as quickly as I could. Without that job, it would have been exceedingly difficult to have maintained my newfound optimism, and I might very well have drifted back into a prolonged depression.

Thanks to strong guidance from my macroself, I did indeed find the employment that was the best thing for me. To begin my job search, I had made a list of several employment agencies to visit. When I arrived at the first agency on my list, however, I found myself unable to enter through their doors. There was no overt reason why I should be reluctant to enter, yet something told me not to go through with it. Yielding to this inner voice, but quite confused, I departed those premises and headed to the next place on the list.

The next agency "felt" good the moment I entered. Despite my extremely unimpressive work history, the agency agreed to provide assistance in finding the right job—with no fee required of me. Within a day or two, I interviewed with an accounting firm in the bustling Loop. In another couple of days, I was hired as a copy editor/proofreader at a perfectly adequate salary. Within another week or so, I had rented a pleasant studio apartment in a classy residential area adjacent to Lake Michigan. My life changed drastically for the better because I had not gone through the doors of the first agency.

I was to remain with this accounting firm for nearly five years. I greatly enjoyed the company of the people with whom I worked, and the requirements of the work helped me to sharpen my English skills. Because this was a very large firm, I was able to transfer to their Los Angeles office when a very strong desire arose in me to relocate to the fresh horizons of California. In the Los Angeles office, I also enjoyed wonderful social relationships with the people I worked with, and I could not have been in a more congenial situation.

A religious person would say that "God" guided me to the right job situation. The same God who is busy causing floods, famine, and earthquakes in various locales around the globe is apparently also very interested in seeing that I get hired by a nice accounting firm. The concept of the macroself is much more reasonable than the idea of an omnipotent God. The macroself is the extension of the self; therefore, it would have every reason to care about what sort of job I found.

Scientific fundamentalists would have us believe that we never receive any guidance at all—that this inner voice that we believe we hear is just our imagination. They would say that it was just a lucky thing that I went to the right agency at just the right time—nothing but dumb luck. All true religious believers know, for a fact, that we can be the recipient of guidance "from above." However, God Almighty is not talking to us; it is our macroself.

It should also be reiterated that the macroself does not operate from a position of absolute benevolence. The macroself is not an angel on our shoulder. The macroself guides us in a way that is likely to be the most creatively satisfying path to follow. Sometimes, this path will be an utterly delightful one where we get to reap all sorts of wonderful rewards. Other times, this will not be the case.

Many years later, under very different circumstances, I found myself back in Chicago and again looking for a job. This time, nothing went well at all. I experienced complete confusion as to what direction I should take. I had no sense of guidance whatsoever. As it turned out, it was best for me not to find a job in Chicago at that time. A weekend trip to my hometown in Michigan turned into a one-way trip when I found out my mother had suddenly died. A job in Chicago would have served no purpose at all, as I was to discover.

While I was a resident of Chicago in the earlier period, I became a member of a very fine Presbyterian church. This church had many things to recommend it. It had beautiful Gothic architecture, an eloquent preacher, an excellent singles ministry, and was located very near my apartment. This church became a major part of my life for about two years, and it demonstrated to me in a very significant and personal way all the very good things that religion can offer people. My intellectual commitment to Christianity was never strong, but my emotional enjoyment of church life was very real. When I moved to a different neighborhood in Chicago, I ceased to attend that church regularly and never found another church to attend that had so much to offer. Eventually, I reached the point where I completely lost interest in church life. But I never forgot how good it could be.

Karl Marx labeled religion "the opiate of the masses," but he was wrong. Religions are the support systems of the masses, and this is not the same thing as an opiate. Whether it be a church, temple, mosque, or open field, religious institutions can offer a helpful, life-affirming solidarity for many people that might otherwise be missing from their lives. The warm, nurturing spirit provided by the religious life should never be denigrated. While theology deserves to be denigrated, emotional support should never be.

During the time that I considered myself at least a nominal Christian, I made a point of reading the Bible thoroughly in an attempt to learn what about it had inspired so many people. I read the laborious Old Testament once—every word of it—and the New Testament three times, in three different versions. Although I had the expectation that I would be inspired and enriched by reading the Bible, this was certainly not the case. I believe that the Bible, and all other major religious texts of all the world's major religions, has been invested with genuine creative power by the macrocreation force. Yet this creative power is not, in any way, universal.

The fact that the Bible had no impact on me does not diminish the legitimacy of its impact on other people, and this would certainly hold true for the Torah, the Koran, and other "sacred" texts. When a religious text has a significant impact on a person, this impact needs to be considered absolutely "real" by those who are not impacted at all. The Bible and other religious works have transformed countless lives for thousands of years. This is a very real phenomenon and should not be considered suspect by scientific rationalists simply because the related theologies are themselves completely invalid. All theologies are invalid, yet all religious experiences are quite valid. This is not a conundrum.

After relocating to California, I had a brief but meaningful "affair" with the Mormon church. There were several things about the Mormons that intrigued me. First, I very much appreciated the Mormon emphasis on community—It is a very supportive church, and each member is considered to be very important. In the Mormon church, you cannot just get "lost in the crowd," and this was important to me, since I had very little social support in my life. Secondly, after visiting their temple in Los Angeles, I became interested in the mystery and secretiveness that the church fosters. This aspect of their church life captured my imagination. Third, I read and became fascinated by the Book of Mormon.

For me, as a reading experience, the Book of Mormon was everything the Bible had not been. I read it in about a week, which is a pretty fast rate for such a demanding text. I definitely felt the creative power that this religious tome had to offer. Shortly after completing the book, I formally joined the church.

Soon, however, I became aware of the conformist nature of Mormon life. Independent intellectual inquiry was most certainly not encouraged. It felt to me that questioning doctrine, in any way, was not permitted, and this was something that I could not abide. The church asked for unquestioning obedience, and this was something I could not provide. After a couple months of membership, it was over.

Despite my genuine appreciation for the good things that church life had to offer, I felt I had no choice but to leave religion behind. My search for an intellectual church could not be confined by the limits that any church had as part of its "package." I needed the social support that church life had to offer, but I could never be so insincere as to pretend to buy into a particular theology. While the Unitarians offered a very open environment, they appeared to be a group that was in no hurry to actually find any answers to any of their questions, and I believed that answers were possible to find and "nail down," at least on a theoretical basis. As far as I am concerned, a faith-based support system has more value to a human being than theoretical truth, yet I was on a quest that could not be turned back, no matter what the consequences to me.

In the mid-1980s, I began reading the writings of the eighteenth century Swedish mystic Emanuel Swedenborg. From the beginning, I felt that he had some useful metaphysical insights to offer. Over the course of several years, I read several thousand pages of his fairly dense prose, looking for any nuggets of philosophical truth that I could find.

Swedenborg was a respected man in his time, at least in Sweden, and his ideas were taken seriously by society at the time. After his death, a small church formed to carry on his philosophy. In the Los Angeles area, I visited one of the Swedenborgian churches and was impressed by the fervent sincerity of its members. But, by this time, I was growing increasingly disenchanted with Swedenborg's writings, and so I did not become involved with this particular faith.

Eventually, I had reached the conclusion that Emanuel Swedenborg had "gone off the deep end." His monastic lifestyle that stressed self-denial in all earthly pleasures helped to fuel his grandiose tendencies. Ultimately, his vainglory got the better of him, and he went insane. While this man had noble intentions with his philosophical work, he needed to magnify the value of his philosophical accomplishments in order to justify all the satisfactions of life that he had forsworn. This is a common mistake that mystics make. It is absolutely necessary to keep one's feet planted firmly on the ground if one is to attempt to soar to philosophical heights.

Swedenborg's failings proved to be a significant lesson to me as I conducted my own philosophical quest. It was clear to me that

Swedenborg felt that he was impervious to self-delusion, and this was enough to doom him. It was clear to me that self-delusion is always lurking in the bushes for someone who has set out to find metaphysical truth, and I vowed that I would never lose sight of that fact. It was so abundantly clear to me where Swedenborg had gone wrong that I knew his mistakes would help me to go right.

In the coming years, as I continued to search for insight, I made it a point to not go overboard in anything that I did. I knew that it was important to never make any demands—if insight were to come to me, it would come in whatever amount of time was desirable, if it were ever to come at all. I always wanted to know more than I did about the nature of existence, but I knew that I could not insist on ever knowing anything more. Each insight that came to me could prove to be the last insight that would ever come, and I had to accept this possibility. All I could do was to remain open to the occasional glimpse of metaphysical "truth" that might come my way. In the meantime, I tried to live my life in a satisfying way and keep philosophy in soft focus, idling quietly in the background.

After finishing with Swedenborg, I shifted to the books of Carlos Castaneda. It was back in college in Michigan in 1975 that I had first started reading him. His first four books were truly mesmerizing. For me, they possessed more creative power than any books I had ever read or ever would read. Those books changed me and changed the course of my life. Their effect was so powerful on me that it was rather frightening and, after college, I put them aside. Almost ten years passed before I was ready to consider them again. With great eagerness, I renewed my association with Carlos.

There can be no doubt that Castaneda is a controversial figure. His claims regarding his anthropological research are quite questionable. There is every likelihood that his accounts relating his adventures with his teacher Don Juan are heavily fictionalized. If Carlos Castaneda had presented himself to the world as a literary philosopher rather than as a serious student of anthropology, his reputation today would be significantly enhanced. Of course, the fact that Castaneda engaged in the use of hallucinogenic drugs brought him into direct conflict with society's reality police, who could be counted on to shoot him down with whatever means were at their disposal. Unfortunately, Castaneda himself, through his vain attempts to appease the anthropological establishment, brought a world of difficulties into his life and muddied his own reputation.

As far as I am concerned, Carlos Castaneda is the greatest philosophical storyteller the world has ever seen. He managed to achieve a very difficult thing, to transcend the human-centered perspective, thus allowing him to scale greater philosophical heights than anyone who came before him. He realized that human beings do not inhabit the center of all existence. He knew that at the center was an abstract core that was as nonhuman as anything could be. This point of view marked a significant advance in the science of philosophy.

Unlike the typical existentialist, Castaneda knew that there are many realities that we can inhabit, places that are as "real" as our usual waking reality. He recognized the symbolic and imaginative nature of these realities, and he had glimpses of the creative nature of existence. Castaneda, it can be said, was an imaginative philosopher—a poet, an artist, a mythmaker. His fictionalized philosophy brought out the fictive nature of existence itself. Those who think of themselves as rationalists and whose dispute with Carlos Castaneda is rooted in the mundane— Castaneda's "lies"—miss the point of everything that Carlos and Don Juan had to say in all the many books that were written. In a fictive reality, a "lie" is simply the creative expression of existence itself. To put it another way, a poet cannot lie.

Although I am very impressed with his achievements, I have attempted to go in a very different direction than Carlos Castaneda. My goal has been to make the macrocreation theory, as well as the macrocreation philosophy, as utterly unimaginative as possible. While I have indulged myself in a fair amount of speculation, I have attempted to keep my speculation within the narrow confines of reasonable logical analysis. Castaneda achieved the ultimate in imaginative philosophical speculation, and there is no one else who can inhabit the worlds that he created. My goal was to stay very close to the ground and forgo the merits of art.

In 1985, shortly before I moved away from California, I had a near-death experience that was not completely typical for this type of thing. As far as I know, I was not ill and not dying. I was asleep and dreaming, but it certainly was not an ordinary dreaming experience. I believed myself to be dying, and I traveled through the funnel (as opposed to a tunnel) that led to the "other side." Even though I truly felt to be involved in the dying process, there was no fear whatsoever.

At the far end of the funnel, I found myself in a place that did not correspond with the standard near-death accounts. There were no "dead" friends or relatives to greet me, nor were there angels and heavenly gates. Instead, I found myself on a lush tropical beach. On this beach, there was a pack of magnificent, peaceful lions. Although I did not fear these animals, I did respect their innate power and dignity.

This scene lasted just a short while, then I woke up in my own bed. However, I was not back to "normal." I found myself as clear as crystal, with not a single thought in my head and temporarily incapable of forming one. After awhile, my usual thought processes returned, and all was well. But this was a near-death experience that has no ready explanation. Since ill health was presumably not involved, then there had to be another function being served, but I really have no useful explanation to provide.

Many people cherish the typical near-death experience, because it sounds so wonderfully appealing to be surrounded by the white light of blissful love and to be greeted by all your departed friends and family. Since reports like these are so common, there is no doubt that they frequently occur, but there is no reason to believe that the near-death event has to follow this happy scenario. Most likely, there is an infinite number of near-death events that can take place, with my lions on the beach scene being just one option. Not all "heavenly" visits will feature religious trappings and invocations. Religion exists to support humans in their human lives. On the "other side," there is no need for religion. Thus, for many new arrivals, religious ornamentation will be nonexistent, right from the very start.

Shortly after my near-death dream, I departed California in a major rush. I sold or gave away virtually all my earthly possessions. An extremely strong internal urging compelled me to leave the West Coast and go back East. The energy of Los Angeles felt profoundly incompatible—it was quite intolerable, and I felt that I had no choice but to flee. My plan was to spend several weeks in Michigan as a vacation, then go on to New York City and explore that area. When my plane landed in Michigan, I felt enormous relief. My mother welcomed me and my cat into her home, and we settled in for a delightful time in the lovely October autumn atmosphere.

My mother and I enjoyed each other's company for the several weeks that we were together. Although I was focused on the upcoming trip to New York, I also appreciated the charms of my old home territory in Michigan. It felt quite gloriously wonderful to be out of California. In early November, my mother and I went to Toledo, Ohio to visit her favorite uncle. After the visit, she took me to Toledo's airport, and I boarded a plane for New York City. This would be the last time that I would ever see her.

After spending some time in New York, I went to Chicago and was considering locating there instead. At the time of her death in a traffic accident, she was the only one who knew where I was. It was not until I returned to Michigan for Thanksgiving that I was informed of her death. There is no question in my mind that my extremely hasty departure from California, and my long visit to Michigan occurred as a result of the instigation of my macroself. My mother and I were given a wonderful opportunity to spend some final time together, but only because I yielded to the promptings of the macroself, which had access to information that my human self could not possibly have. It was not the voice of God who spoke to me, nor a guardian angel; it was the part of myself that exists at the macro level of reality.

When I had departed California in a mad rush, I had divested myself of nearly all my possessions. When my mother died, I acquired all of hers, which certainly worked out very nicely for me. I remained in Michigan for about two and a half years. While I enjoyed the scenery and the pleasant sentimentality of the place, I eventually began to feel somewhat understimulated, and I hoped to find a new path in life. When it came, this new path took me to Austin, Texas.

I took a position as assistant to the director of the Austin Seth Center, an educational, nonprofit organization that promoted the work of Jane Roberts. Jane and her husband Robert produced a series of books featuring the philosophical beliefs of Jane's channeled personality, Seth. I had read these books while living in Michigan and had been impressed with some of the ideas that were contained within. For instance, Seth talked about the concept of "awareized energy," which certainly fit in with my own thinking. I did have some reservations about the totality of what Seth had to say, as well as the channeling phenomenon in general, but I felt going to Austin and helping out the director in her duties would be an interesting use of my time, and so I performed my duties with a certain measure of enthusiasm. Texas was a new and intriguing experience, and I was in the mood for new things.

Jane and Rob considered themselves rational thinkers, and they always attempted to subject the material that they received from Seth to some degree of intellectual analysis. They shared some of the material with physicists and other scientists and found some measure of appreciation in their ranks. Yale University was sufficiently impressed with their output to accept the collected papers for their archives. The Seth material represented a step away from the usual mumbo jumbo that had dominated metaphysical writings in the past. Nonetheless, I eventually found myself to be quite disenchanted with the overall nature of the work that Jane, Rob, and Seth produced.

Ultimately, after much contemplation, I decided that the Sethian philosophy lacked any true coherence. The ideas were actually rather a jumble—formless and incoherent—they sounded intelligent and "scientific" unless examined with sufficient discrimination. Seth went round and round in circles, never really getting anywhere. It became clear to me that the Seth philosophy was the product of limitless imagination, and logic played little role in his thinking.

While at the Austin Seth Center, I did my best to support the work of the center, but my increasing disillusionment made it quite a strain. For the year that I was there, I did a lot of useful things that were of help to the director, a charming elderly lady, and I cannot regret my time devoted to the center, but I could not continue on with any genuine sincerity.

The final act of my Sethian period took place in upstate New York. After leaving Austin, I had spent a year in Dallas pursuing the radio business in a fitful manner. When I quit the little radio station where I was working, I decided that I wanted to leave Texas altogether at that time and opted for Rochester, New York, just because I thought I might like it. Not far from Rochester was Elmira, where Jane and Rob had lived. I went down there to take some pictures of the house where they lived to send back to the director of the Seth Center, who had never been able to make the trip.

By this time Jane was dead—her final years had been spent in a local hospital, suffering from severe rheumatoid arthritis. She had continued to channel Seth during this period, and nothing he said appeared to be of any value to her. Despite this, her husband, Rob, had not turned his back on the Seth material. I met him in the house that they had shared and took a lot of photographs to send back to Austin.

Even with the strong misgivings that I had, it was a somber thrill to be in the locale where so much metaphysical material had been produced. Coherent philosophical truth had eluded them, but Jane and Rob had made a noble effort.

As a general rule, information that is acquired using a channeled, disassociated personality will have little intrinsic merit. Simply because a personality has its focus in the macro level of reality is no reason to be impressed with the "wisdom" that it imparts. There is never a reason to opt to disassociate oneself from one's own personality—we are innately linked to our own macroself, and no special effort need be made, apart from some gentle meditation and mild contemplation, to seek out insights from the macro level. Traditional channeling can be seen as a histrionic performance, with no special merit accruing.

While still living in Austin, I participated in a session of past-life regression that yielded some interesting results. While I am not hypnotizable, I did achieve a sufficiently relaxed state that did allow for some imaginative scenes to be played out in my mind's eye.

I saw myself as a painter living in England in the early part of the twentieth century. This man was married to another painter with whom he was very competitive. Because of his deep-rooted insecurity, his marriage was marked by fits of corrosive envy and jealousy. He had a bitter spirit, and he brutalized her with his words. The marriage may have very well ended in tragedy.

When I had visited London as a teenager, I was surprised by how uninteresting the city seemed to me. While I had no particular sense of déjà vu, all the street scenes that I beheld felt quite stale. In contrast, the other nations of Europe that I visited on that trip felt fairly fresh and invigorating. All of these impressions are certainly quite subjective, but they do fit in with this "associated life" of a painter that I had seen in the regression.

While traditional ideas regarding reincarnation have no appeal in terms of logic, the concept of an energy-related associated self is more appealing on a rational basis. Since energy exists as a continuum, there is every reason to believe that the individualized energy matrix that comprises oneself would have a direct relationship, within the greater energy continuum, to another self. This relationship would be very intimate and meaningful, so much so that people who have "reincarnational" memories confuse those other lives with their lives. I do not believe for a moment that "I" lived in England in the early twentieth century, but I can believe that I have a strong energy relationship to someone who did.

One of the most disturbing events of my life took place while I was living in Rochester, New York. One summer, I decided to go to Lily Dale, a small community of psychics and mediums that was located south of Buffalo. The drive there took me through gorgeous, hilly countryside, and the charming village of Lily Dale itself was a delight to see. There was a lecture I planned to attend, then I thought I would walk around the village and see if I felt like getting a reading from one of the resident psychics.

After leaving the lecture, I walked up and down every street, looking at the shingles that hung outside each cottage, trying to decide if I wanted to get a reading or not. In the past, I had been generally unimpressed with the quality of the readings that I had received, so I felt some genuine reluctance in laying down some cash for another one now. Finally, I decided to save my money and got in my car and drove home.

On the drive back to Rochester, I began to have the feeling, quite subtle at first, that there were thoughts in my head that were not my own. These thoughts were not disturbing for their content, which was innocuous, but because they were there at all. By the time I arrived home, I was becoming rather agitated with these uneasy sensations. It was as if I had a crowd of tourists in my head.

Not knowing what else to do, I went to bed. As I lay there, the cacophony of whispered thoughts became utterly intolerable. Summoning the full force of my intent, I informed my macroself that I had reached the limits of my endurance. I did not plead or beg; I simply informed my macroself that I could not bear it anymore and that it had to stop. In an instant, there was the blessing of peaceful silence.

Apparently, while strolling around the charming village of Lily Dale, I had picked up a number of personalities who were floating around in the highly charged psychic atmosphere. These bits of personalized energy clung to me and made the journey with me back to Rochester. They presumably had no malicious intent. When it became clear to them that their presence was not welcome inside my head, they departed without a fuss. Needless to say, I made a promise to myself never to return to Lily Dale.

I have to conclude that this event occurred to give me a vivid lesson in the nature of free-floating personal energy. These personalities can surround us and, generally, we will take no notice. Because the Lily Dale environment is fairly unique, I had a fairly unique experience. Fortunately, my macroself paid attention to my demand that this experience come to a quick end. When it comes to pain, whether it be "physical" or psychological, the relationship between the human self and the macroself is a very direct and useful one.

I have had many personal experiences with various types of pain that have influenced my overall philosophy on the subject. One notable incident occurred in Texas, where I experienced an excruciating attack of back pain. Ordinarily, back pain was not something that I had trouble with, so the incident was quite disturbing.

The pain began in the morning then built steadily during the day. By late afternoon, after I had returned home from the Austin Seth Center, the pain had escalated to the point where I could barely move. There was no position that I could take that did not cause pain. I attempted to walk to a convenience store about a block away for some pain reliever, but I had to turn back.

In desperation, I decided to summon my intent in order to alter the dynamic of the situation. I chose, at a very deep level, to not accept the reality of this pain in my life. Whatever "medical" explanation that might exist to explain the pain was irrelevant. My choice was to dispel the pain the same way I would opt to wake up from a bad dream.

In order to manifest the force of my intent, I began to move my body in all sorts of ways, the kind of movement that would have, presumably, resulted in extreme pain. I had to demonstrate to myself that the power of the pain could be dealt with summarily and dismissed from my reality.

Within a minute of taking this action, the pain had completely dissipated. I was completely free of pain, and it truly seemed if the reality of the pain had been dispelled like a fantasy. Even more wonderful, this extreme degree of back pain never entered my life again. A battle had taken place between me and the pain, and I had been victorious.

In order to experience pain, one must accept that pain. For complex psychological reasons, all of us can learn to tolerate chronic, low-grade pain in our lives. This type of pain can play a meaningful role with significant creative ramifications. However, bouts of extreme pain are a very different matter. This kind of pain can manifest itself in a way that is fundamentally intolerable. This is the type of pain that should be rejected when it reaches that level of unacceptability. Summoning one's intent, at full force, informs the macroself that a critical juncture has been reached. The personal power of the human self is sufficiently strong that the macroself will pay attention.

Many years later, while suffering through several bouts of severe dental pain, I was again able to successfully use my intent to strongly influence the level of pain that I was experiencing. Much of the time, when the prescribed pain medication was working fairly well, I could accommodate myself to the discomfort. Every now and then, however, it was necessary to "get tough" with my macroself.

In one instance, I was able to dispel the pain using the force of my intent for about twenty-four hours. When the pain returned, it was not nearly so strong. Other times, the focus of my intent served as a potent "pain pill" that did its job very well. It was a great relief to me to understand that I was not helpless in these situations.

During this period of dental problems, I never attempted to focus my personal power unless the degree of pain that I was experiencing was truly severe. The emotion of anger can be helpful in these situations, and it is a far cry from making a timid little plea to an omnipotent God to please help me. Most of the time, we can cope with the incidents of pain that come into our lives—pain is to be expected from time to time. In some instances, however, the level of pain that we experience will be beyond our ability to cope. This is when we need to take action.

Turning to the realm of precognitive endeavor, it is clear that it is quite difficult to reliably predict future events, no matter how talented a psychic individual may be. This may be because the nature of time is very fluid and subject to infinite changes, or it may be because a high degree of psychic ability is simply not to be found in the human range of gifts, or possibly both. However, it is clear that, upon occasion, a human being is fully capable of sensing a future event before it occurs in our reality. While a scientific fundamentalist would deny this is possible, this is simply due to the fact that they have no intellectual maneuvering room. They are quite boxed in by their rigid psychological makeup. Most people realize that any one of us can have an occasional glimpse into the future, even while admitting that this ability is not entirely reliable.

While I do not consider myself to have received the gift of precognition, there have been occasions when I have felt in tune with future events. I know that if I attempted to predict the future on a routine basis, I would almost certainly fail miserably, as do most "professional" psychics. Obtaining a genuine glimpse into the future occurs as the result of a mysterious process, and perhaps it will always remain that way. Having a reliable precognitive ability is most likely not in humanity's best interest.

As a young person interested in the world of American politics, I found myself especially fascinated by the great scandal known as Watergate. I wondered what the outcome of the scandal would be for Richard Nixon. Over a year before Nixon's resignation, I already knew, somehow, that this unprecedented event would occur. Four months before the resignation, I knew when it would occur. These insights did not represent guesses; they represented knowledge. I had a genuine sense of certainty regarding the future.

When the leader of the Soviet Communist Party, Andropov, died, the name of Mikhail Gorbachev was one of many that was mentioned as a possible successor. The first time that I heard the name Gorbachev, I knew not only that he would someday lead the Soviet Union, but that he would be a very significant leader, far different from the colorless Andropov or the ineffectual Brezhnev. As it turned out, Gorbachev had not yet achieved sufficient importance to be able to succeed Andropov, but his time did come later.

Predicting the results of American presidential elections is something that can often be fairly easy to do for almost anyone—accurate pre-election polling being what it is. But the 2000 presidential election was something a little different for me. I knew, without any doubt, before George W. Bush had earned a single vote in a caucus or a primary, that he would become the next president of the United States. This outcome held no particular pleasure for me, but I could readily understand how his election would be the right creative choice for that moment in history. Politics, like all things, is a creative affair.

September 11, 2001 impacted me in a personal way before that day had actually dawned. On the evening of September 10th, I began to feel increasingly edgy—it can be likened to the way that barnyard animals give indication of an impending storm or earthquake by their agitated behavior. That night, I could not sleep at all. When morning came, I felt so very frazzled and uneasy that I called in "sick" to work, something that I rarely did. The whole experience was leaving me quite perplexed. I cannot say that my thoughts turned to New York City or to Washington, but I did feel something was unfolding.

It was my routine to avoid listening to the news of the day until the evening television newscast but, that morning, I turned on the radio at the very time that the first tower of the World Trade Center was crumbling to the ground. Like everyone else, I followed the dramatic events of the day as they played themselves out. There was certainly no doubt in my mind that my macroself had influenced my emotional state, so that I would choose not to go to work that day and thus be able to focus on the important events that were then occurring.

As a general rule, I do not make any particular effort to predict future events. One of the pleasures of this earthly existence is the element of surprise. Since I very seldom gamble for money, knowing the outcome of an election or a scandal before it happens has little practical value. An occasional glimpse into the future is a pleasant diversion, while a continuous foresight would probably prove to be something of a burden.

Some people have experienced precognitive dreams, but I never have. However, my personal dreaming adventures have been so vivid, vast, enlightening, and entertaining that I certainly have a full appreciation for them. Those people who have very little recall of their dreams do not know what excitement they are missing. Understanding the world of dreams is essential to comprehending the nature of reality.

I know that the dream environment is a "physical" world because I have experienced this physicality countless times. Some of my most interesting dreams have taken place in a variety of urban landscapes where, many times, I have been able to take note of the richness of detail in the cities that surrounded me. With the warmth of the "sun" on my back or the chill of "winter" in my bones, I have walked down street after street, gazing at building after building. This reality is as real as real can be.

It appears that my consciousness is never entirely submerged into the dream experience. Although I always recognize the world around me as being "real," I nonetheless realize that it is a fluid reality and that this gives me options. For instance, I never have to suffer though a full nightmare experience because I always exit the dream the moment that it becomes more intense or unpleasant than I am willing to endure. This exit strategy is always with me, even if I remained unaware of it until the very moment that I required it.

On some occasions, the level of my waking awareness will exactly match my dream awareness. There have been times where my thought process has been utterly continuous from the dream state to the waking state. In these instances, the transition from being asleep to being awake is so smooth as to be virtually nonexistent. This points out the fact that all realities can be considered dream realities—Reality functions as a dream functions.

On a few unpleasant occasions, I have "awakened" to find myself back in my own bed in my own bedroom, yet I am not quite back in my own reality. Generally, I begin to realize that something is amiss when I begin to float toward the ceiling. This alerts me to the fact that, whatever reality I am in, it is not quite the correct one. Realizing this causes me to wake up again and, this time, I have made it all the way back. Our waking reality rests on a narrow beam of focused attention, and it is not that terribly difficult to slide off.

Once in a great while, I return from a dream experience that took me so "far away" from my usual waking reality that I find myself rather disoriented upon my return. I will have no sense of how long I have been asleep—it might have been for twelve minutes or for twelve hours.

It will take a moment or two to reassemble the pieces of my waking self and get back fully in the swing of things. Possibly, this type of experience represents a rather abrupt awakening from a "macrodream," a dream that takes place in macroreality. I can never remember the nature of the dream that I awoke from, which is a characteristic of the macrodream.

From personal experiences, I know that it is possible to dream even while one is awake. Of course, it is necessary to be very relaxed, but it is not necessary to be asleep in order to dream. One can bring an image to the mind's eye and then let that image evolve into a brief, fragmentary dream. This waking dream cannot usually be sustained for more than a few seconds, unless I have made the effort to do more than that. This little ability that we have shows us once again how our reality is constructed—through the imagination.

Our dreams have more to tell us about the nature of reality than any physics textbook. All the clues we need to determine the nature of our existence have been right before us all the time—awake and asleep.

Concluding Remarks

For most of my life, I have been on a quest for wisdom and enlightenment, but I never assumed, not for a moment, that this quest would be a successful one. When useful insights came to me, they came slowly, and quite irregularly. Never did I have any certainty that one insight would necessarily be followed by another. While I definitely desired to learn a lot about the nature of existence, I was prepared for the possibility that I would learn only a little. My expectations were as modest as I could make them.

In addition, my claims regarding the macrocreation theory are as modest as I can make them. I cannot make the claim, on any rational basis, that the macrocreation theory represents metaphysical truth. The only people who can make the claim to know what the truth is are people who have a psychological need to make this claim. I hope that the introductory material in this book makes it quite clear that every statement in this book is speculative in nature. It has been important to me that the ideas that comprise the macrocreation theory and the macrocreation philosophy be expressed clearly enough so that they can be understood and taken seriously on their merits, but it is not important to me to change hearts and minds. Evangelism is not an aspect of my personal character.

Someday, I believe, the religions of the world will have run their creative course and will no longer meet the needs of very many people. In the future, a religion that now holds the loyalty of hundreds of millions of people may be reduced to the size of the present-day Druids. It may also be the case that scientific atheism may have already peaked in its popularity, since it combines irrationality with inhumanity, which should have a rather limited appeal. Theoretical physicists have been chasing their tails for a long time and presumably will continue to do so into the indefinite future, since this appears to be an activity that they enjoy. Considering all this, there may, possibly, be a place for the philosophy of macrocreation.

This world will be a better place when there is better comprehension of this world. Ignorance is not bliss—it is confusion. While it can reasonably be said that knowledge is power, it can also be said that comprehension is the ultimate power.

I believe that forming energy connections with each other is a vital function of this earth reality. This is why it makes perfect sense to love one another. By strengthening our bonds with each other, we strengthen the fabric of reality, and of existence itself.

In conclusion, I believe that one statement that needs to be made is this: psychology rules the intellect. For all of us, the belief system that we embrace will be the one that has the greatest psychological appeal. There is simply no escaping this. We tend to believe what we believe because it feels good and it feels right. This is one fundamental reason why I cannot function as an evangelist. Our belief systems must support us emotionally. Genuine rationality has to take second place to this. This is how it must be.

About the Author

At a family gathering when the author was six years old, an ouija board was placed in front of him and he was told that he could ask the board any question that he wanted. The boy pondered for a moment, then arrived at this metaphysical brain-teaser:

"How was God born?"

The board, of course, did not provide a satisfactory answer. But this fundamental question lingered. As the boy grew into a man, he sought information from science, religion and New Age thinkers. Throughout it all, he maintained an open mind.

Many years elapsed. Very gradually, an original belief system emerged focusing on the creative qualities of energy. This is the book that finally answers the question.

www.ingramcontent.com/pod-product-compliance
Lightning Source LLC
Chambersburg PA
CBHW032007170526
45157CB00002B/576